Introduction to
Atomic and Molecular Structure

Introduction to
Atomic and Molecular Structure

JACK BARRETT

Lecturer in Inorganic Chemistry,
Chelsea College, London University, England

JOHN WILEY & SONS LTD.

LONDON   NEW YORK   SYDNEY   TORONTO

Copyright © 1970 John Wiley & Sons Ltd., All Rights Reserved. No part of this publication may be reproduced, stored in a retrieval system, or transmitted, in any form or by any means, electronic, mechanical photocopying, recording or otherwise, without the prior written permission of the Copyright owner.

Library of Congress Catalog Card No. 74-93558

SBN 471 05416 X Cloth bound
SBN 471 05417 8 Paper bound

Set on Monophoto Filmsetter and printed by
J. W. Arrowsmith Ltd., Bristol, England

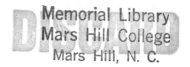

Preface

Since the early nineteenth century atomic and molecular theories have developed at ever increasing rates. The situation today is such that an undergraduate could be fully occupied in dealing with this subject alone. This is clearly not possible, so the student who wishes to become a theoretical chemist has to learn his trade at the post-graduate level. For a broadly based qualification in chemistry the student requires a general understanding of the aims, techniques and successes of chemical theory. He can then use this understanding to deal with the many and varied facts which are known about elements and compounds. It is the purpose of this book to supply a basis for this understanding.

For educational and economic reasons the treatment is by no means a comprehensive one, but the author has included mathematical derivations that are indispensable for a fuller understanding of the subject. It is certainly possible to write a book on the subject which does not include mathematics. Such a book would contain the bare statements of the results of the mathematical treatment of chemical theory. In the author's experience this is very unsatisfactory for the enquiring student whose level of acceptance lies beyond a non-mathematical approach where none of the fundamental relationships are derived, and so in trying to give a general account of chemical theory, he has considered it not possible to dispense completely with the mathematical aspects. Even so the treatment remains a qualitative one.

This book has developed from courses given to undergraduates at Chelsea College in their first and second year course units. It is hoped that the book will provide first and second year students with a satisfactory basis for the understanding of theoretical chemistry. In addition it should be of use to sixth-form teachers who have to prepare their pupils for university entrance.

I wish to thank several people for their assistance to me in writing this book.

Acknowledgements are due to John Barnes, Godfrey Beddard and Tony Beezer for reading the book at the manuscript stage and for supplying helpful criticism. The book was also read in its early stages by two anonymous referees whose comments were very useful. I thank them for their conscientious criticism of the material.

Jeffrey Arnold was responsible for the original drawings of the many diagrams and Figures which appear in the text and I am extremely grateful to him for the excellent job he did.

I thank Janet Barrow for her extremely efficient transformation of the hand written book into the typed manuscript.

Finally I thank Tony Beezer for his very welcome assistance in reading the proofs and in the preparation of the index.

JACK BARRETT 1969

Contents

Introduction

    The Building-up of Matter xiii
    The Approach to Theoretical Considerations . . . xiii
    Units xiv

1. The Quantization of Electromagnetic Radiation

    1.1 The Electromagnetic Spectrum 1
    1.2 The Nature of Electromagnetic Radiation . . . 3
    1.3 Energy Distribution in Black-body Radiation . . 3
    1.4 The Photoelectric Effect 6
    1.5 The Compton Effect 10
    1.6 Duality of Radiation Theories 11
    1.7 Summary 12
    Problems 13

2. The Quantization of Electron Energies in Atoms and Molecules

    2.1 The Quantization of Electron Energy Levels . . . 14
    2.2 Hydrogen Emission Spectrum 14
    2.3 Emission Spectra of Elements other than Hydrogen . 20
    2.4 Molecular Absorption Spectra 21
    2.5 Ionization Potential Data for Atoms 26
    2.6 Ionization Potential Data for Molecules . . . 33
    Problems 35

3. The Diffraction of Electrons and the Uncertainty Principle

    3.1 Electron Diffraction 37
    3.2 Matter Waves 39
    3.3 The Observation of Electrons and the Uncertainty Principle 41
    3.4 Consequences of the Uncertainty Principle . . . 45
    Problems 45

4. Quantum Mechanics

4.1	Energy Levels in the Hydrogen Atom	46
4.2	The Schrodinger Equation	47
4.3	Interpretations of ψ	51
4.4	Limitations to the Nature and Form of ψ	52
4.5	Solutions of the Wave Equation for an Electron in a One-dimensional Potential Well	52
4.6	Normalization of the Eigenfunctions	56
4.7	Degeneracy	56
4.8	Dimensions and Quantum Numbers	60
4.9	Non-zero Values of the Quantum Numbers	60
4.10	Conclusions	61
	Problems	61

5. The Application of Quantum Mechanics to Atomic Problems

5.1	The Eigenfunctions and Eigenvalues for Hydrogen-like Atoms	62
5.2	Atomic Orbitals	64
5.3	The Eigenvalues of the Hydrogen-like Atomic Orbitals	72
5.4	Polyelectronic Atoms	73
5.5	Orbital Penetration Effects	76
5.6	Summary	79
	Problems	79

6. The Periodic Classification of the Elements

6.1	Electron Spin and Atomic Emission Spectroscopy	81
6.2	Electron Pairing and Electron Correlation	89
6.3	The Pauli Exclusion Principle and the Construction of the Periodic Classification of the Elements	97
6.4	Summary	109
	Problems	109

7. Combination of Atoms: The Formation of Diatomic Molecules

7.1	Introduction	110
7.2	Chemical Bonds	110
7.3	The Formation of Molecular Orbitals in H_2^+ and H_2	111
7.4	Application of Molecular Orbital Theory to the Homonuclear Diatomic Molecules of the First Short Period Elements	124
7.5	The Hybridization of Atomic Orbitals	131
7.6	The Hydrogen Fluoride Molecule	133

Contents ix

- 7.7 Electronegativity Coefficients 136
- 7.8 Some Other Heteronuclear Diatomic Molecules . . 139
- Problems 143

8. Triatomic Molecules

- 8.1 Triatomic Hydrides, AH_2 145
- 8.2 The Walsh Treatment of Triatomic Hydrides: Symmetry Theory 150
- 8.3 The Sidgewick–Powell Approach to AB_2 Molecules . 173
- 8.4 The Walsh Treatment of AB_2 Molecules . . . 178
- 8.5 The Structures of BAC and HAB Molecules . . . 187
- 8.6 The Molecules, SH_2 and PH_2 189
- 8.7 Summary 189
- Problems 190

9. Other Polyatomic Molecular Systems

- 9.1 The Sidgewick–Powell Treatment of AB_n Molecules where $n \geqslant 3$ 191
- 9.2 The effects of Non-Bonding Pairs 197
- 9.3 Molecules with Multicentre Sigma Bonds . . . 202
- 9.4 Molecules Involving Multiple Bonds 203
- 9.5 Some Inter-related Polyatomic Molecules . . . 209
- 9.6 Some Bridged Molecules 217
- 9.7 Some Organic Systems: Pi Electron Delocalization . . 221
- 9.8 Valence Bond Theory 235
- Problems 237

10. The Solid State and the Bonding in Ionic Compounds and Metals

- 10.1 Introduction: Physical States of Matter . . . 238
- 10.2 Intermolecular Forces 239
- 10.3 The Nature of the Hydrogen Bond 244
- 10.4 Covalent Solids 247
- 10.5 Ionic Solids 249
- 10.6 Theoretical Treatment of Ionic Bonding . . . 252
- 10.7 'Experimental' Treatment of Ionic Bonding: Born–Haber Thermochemical Cycles 255
- 10.8 The Metallic Bond 263
- Problems 268

11. Molecules and Ions Containing Transition Elements

- 11.1 Introduction 271
- 11.2 Crystal Field Theory of Octahedral Complexes . . 278

11.3	Crystal Field Theory of Tetrahedral Complexes	291
11.4	Ligand Field Molecular Orbital Theory	295
11.5	Summary	309
	Problems	

Appendix I Slater's Rules for the Calculation of Effective Nuclear Charge 311

Appendix II The Successive Ionization Potentials (eV) of the Elements of the First Short Period 312

Appendix III The Order of Filling of the Orbitals of some Homonuclear Diatomic Molecules 313

References for Further Reading 317

Answers to Problems 318

Index 321

Introduction

In order to study the chemistry of the elements three fundamental particles are of importance. These are *protons* and *neutrons*, which constitute the *nucleus* of the atom, and the extra-nuclear *electrons*. The physical characteristics of these particles are summarized in Table 1.

Table 1. Characteristics of Chemically Important Fundamental Particles

Particle	Symbol	Mass (gram)	Mass (atomic weight units)	Electrical charge
Proton	p	$1{\cdot}67252 \times 10^{-24}$	$1{\cdot}0072785$	$+e^a$
Neutron	n	$1{\cdot}67482 \times 10^{-24}$	$1{\cdot}0086654$	zero
Electron	e or β^-	$9{\cdot}1091 \times 10^{-28}$	$5{\cdot}48597 \times 10^{-4}$	$-e$

$^a\, e = 4{\cdot}80298 \times 10^{-10}$ e.s.u. $= 0{\cdot}16$ aC.

The proton, in addition to being present in all nuclei, forms the simplest nucleus, that of the hydrogen atom. All other nuclei possess a certain number of neutrons. The most important nuclear characteristic is the *atomic number*, Z, which is the number of protons in the nucleus and is numerically equal to the nuclear charge. This number characterizes the different chemical *elements*, the present range of values being from $Z = 1$ for hydrogen to $Z = 103$ for lawrencium.*

A second term characterizing a given nucleus is the *mass number*, A, which is the nearest whole number to the exact mass, M, of the atom. For example, on the atomic weight scale which has as its basis the mass of the twelve-carbon atom as being exactly twelve (12·00000) the exact mass of the sixteen-oxygen atom is 15·9994. The mass number of the latter is sixteen. The atomic weight of an element is the weighted average of the exact masses of its constituent *isotopes*. For example, the exact mass

---

* Recently element 104 has been reported.

of the thirteen-carbon atom is 13·003354 and this isotope forms 1·107 per cent of the naturally occurring element, the other constituent being the twelve-carbon isotope. The weighted average of the two exact masses is 12·011 which is the *atomic weight* of the element.

It is conventional to place the value of A as a left-hand superscript and the value of Z as a left-hand subscript to the element symbol, e.g. $^{14}_{7}N$, $^{238}_{92}U$. This leaves the right-hand side of the element symbol clear for the conventional chemical indications of ionic charge (top right) and molecularity (bottom right), e.g. I_3^-, or in full $^{127}_{53}I_3^-$.

The difference between the mass number and the atomic number, $A-Z$, is equal to the number of neutrons contained by the nucleus. For any particular value of Z, the value of $A-Z$ is found, usually, to vary over a short range. Consider the nucleus of the element for which $Z = 1$. There are found to be three values for $A-Z$ which are zero, one and two corresponding to nuclei containing one proton and, respectively, none, one and two neutrons. This is an example of *isotopy* and a summary of the *isotopes* of hydrogen is shown in Table 2. Only in this extreme case are the isotopes given different names and chemical symbols. The large differences in mass number cause greater differences in properties than any other set of isotopes. For instance, the rates of the reactions

$$D + CH_4 \rightarrow HD + CH_3$$

and

$$H + CH_4 \rightarrow H_2 + CH_3$$

under identical conditions differ by a factor of six in favour of the formation of HD. The difference in rate is due primarily to the difference in mass number of the abstracting atoms. Isotope effects on the reactions of other elements are observable but are not as obvious as the case already quoted since the ratios of the mass numbers of other isotopes never approach two as with deuterium and hydrogen.

Table 2. The Isotopes of Hydrogen

Name	Symbol	Z	A	$A-Z$ (number of neutrons)
Hydrogen (Protium)	^1_1H	1	1	0
Deuterium	^2_1H (D)	1	2	1
Tritium	^3_1H (T)	1	3	2

Introduction

The Building-up of Matter

Viewed in terms of a constructional process from the fundamental building units outlined in Table 1, the building-up of matter may be dealt with in four large sections, these being classified in terms of the cohesive forces involved and the products of such cohesion. The classification is summarized in Table 3.

Table 3. Classification of Material Structure

Class	Units	Cohesive force	Product
1	Protons, Neutrons	Nuclear	Nuclei
2	Nuclei, Electrons	Atomic	Atoms
3	Atoms	Molecular or Valence	Molecules
4	Molecules	Intermolecular	Aggregations of Molecules

The first class will not be dealt with in detail here except to mention that the nuclear forces overcome proton–proton repulsion to produce existent nuclei. The second class forms the first major section of this work which will be concerned with the construction of *atoms* from nuclei and *electrons*. This section will give a detailed treatment of atomic forces which operate between nuclei and the surrounding electrons. Once these important forces have been treated we shall be in a position to consider the construction of molecular systems, together with an understanding of the chemical reactivity or inertness of atoms in terms of valency forces. This will form the second major section of the book. Finally we can discuss physical states of matter and aggregations of molecules in terms of intermolecular forces.

Although it is expedient to divide material structure into the four classes contained by Table 3, it should be borne in mind that all four types of cohesive force are basically coulombic or electrostatic.

The Approach to Theoretical Considerations

In a largely theoretical treatment of atomic and molecular systems it is very easy to give the false impression that chemistry is becoming a completely theoretical subject. It is hoped that this work will give a sound understanding of the theoretical concepts involved by placing the correct emphasis upon the vast amount of experimental work which has led to the theories. Chemistry is still very much an experimental science and the first part of the book will concentrate on experimental results rather than on the theory, and attempts are made throughout to emphasize the dependence of theory upon experimental observations.

Units

Quantities including mass, length, time, energy, frequency and electric charge must be expressed in terms of well-defined *units*. The units used in this book include those recommended by the International Organization for Standardization. They are known as SI units (Systeme International d'Unites). The basic idea of SI is that a single unit is adopted for a given quantity, and that only decimal multiples and sub-multiples of the unit are to be used. For example, the unit of time is the *second*, and all times are to be expressed in seconds and not in minutes or hours, etc. A summary of the SI units, together with some alternative units, for various quantities is given in Table 4.

Table 4. SI and Other Units in Common Use

Quantity	SI unit	Other units
Time	second (s)	minute, hour, year
Mass	gram (g)	—
Energy	Joule (J)	calorie, erg, electron volt
Frequency	Hertz (Hz)	cycles per second (1 c/s = 1 Hz)
Length	metre (m)	—
Electric charge	Coulomb (C)	e.s.u.

To reduce the numbers required to express a given quantity to a minimum, several decimal multiples and sub-multiples are used. These multiplication factors with their names and symbols are given in Table 5.

In terms of SI units 10^{-9} seconds would be expressed as 1 ns and one thousand grams would be 1 kg.

Whereas times and masses are commonly expressed as seconds and grams respectively, there are cases where the SI unit is not very common compared to usual practice. Energy units are one such example. Common units are the electron volt, the erg and the calorie. The conversion factors for these units and the Joule are given in Table 6.

When describing the energy levels of atoms and molecules it has been usual practice to use the electron volt as the unit. Such energies are expressed as electron volts per atom or molecule (eV.atom^{-1} or

Introduction

Table 5. Multiplication Factors, Their Names and Symbols, As Used for SI Units

Multiplication factor	Name	Symbol
10^{12}	tera	T
10^{9}	giga	G
10^{6}	mega	M
10^{3}	kilo	k
10^{2}	hecto	h
10^{1}	deca	da
10^{-1}	deci	d
10^{-2}	centi	c
10^{-3}	milli	m
10^{-6}	micro	μ
10^{-9}	nano	n
10^{-12}	pico	p
10^{-15}	femto	f
10^{-18}	atto	a

eV.molecule^{-1}). Since there are many experimental situations which involve vast numbers of atoms or molecules it is also customary to express the energies of such assemblies in terms of the kilocalorie . mole^{-1} where the kilocalorie represents 10^3 calories and the mole is $6·0225 \times 10^{23}$ (Avogadro's Number) of the species involved (whether it be an atom, molecule or ion). The relationship between these units is as follows:

$$1 \text{ eV.atom}^{-1} = 23·06 \text{ kcal.mole}^{-1} = 96·7 \text{ kJ.mole}^{-1}.$$

The frequency of an oscillation is usually expressed as the number of oscillations per second (cycles per second), and this is exactly the same in SI units where 1 cycle.sec^{-1} = 1 Hz.

Electric charge is usually expressed in electrostatic units (e.s.u.), the recommended SI unit being the coulomb where $1 \text{ C} = 2·998 \times 10^9$ e.s.u. and represents one ampere-second. The charge on the electron is $4·80298 \times 10^{-10}$ e.s.u. or $1·6 \times 10^{-19}$ C (0·16 aC).

Table 6. Energy Conversion Factors

	Electron volts (eV)	Ergs	Joules (J)	Calories (cal)
1 eV =	1·0	$1·6 \times 10^{-12}$	$1·6 \times 10^{-19}$	$3·83 \times 10^{-18}$
1 erg =	$6·24 \times 10^{11}$	1·0	$1·0 \times 10^{-7}$	$2·39 \times 10^{-8}$
1 J =	$6·24 \times 10^{18}$	$1·0 \times 10^{7}$	1·0	0·239
1 cal =	$2·61 \times 10^{17}$	$4·184 \times 10^{7}$	4·184	1·0

In this book the SI units of time, length, mass and frequency are used without mention of other units. They are in any case very commonly used. The energy units are those which are in common use. Wherever the kcal.mole^{-1} is used the equivalent value in kJ.mole^{-1} is placed in brackets.

1

The Quantization of Electromagnetic Radiation

1.1 The Electromagnetic Spectrum

Electromagnetic radiation is a form of energy which may be classified in terms of the wavelength, λ, or the frequency, v, of a wavemotion such as is shown in Figure 1.1. The frequency of the radiation is related to the wavelength by the equation:

$$v = c/\lambda \qquad (1.1)$$

where c is a universal constant equal to the velocity of electromagnetic radiation *in vacuo* and has a value of $2 \cdot 998 \times 10^8$ m.sec^{-1}. The value of

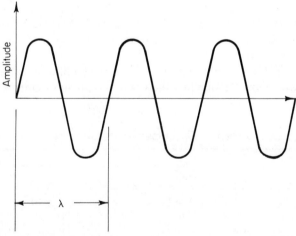

Figure 1.1. A typical wavemotion, showing the wavelength, λ

the velocity of electromagnetic radiation in a medium other than a vacuum is given by c' where

$$c' = c/n \qquad (1.2)$$

n being the refractive index of the medium.

The wavelength of the radiation, λ, is also dependent upon the medium, the value in a medium of refractive index, n, being λ' where

$$\lambda' = \lambda/n \tag{1.3}$$

and λ is the value of the wavelength *in vacuo*.

This being so, it follows that the frequency of the radiation is independent of the medium since we have

$$\nu = c/\lambda \quad (\textit{in vacuo}) \tag{1.4}$$

and

$$\nu' = c'/\lambda' \text{ (in a medium)} \tag{1.5}$$

and, taking into consideration equations (1.2) and (1.3) together with equation (1.5), we have

$$\nu' = c/n \times n/\lambda$$
$$= c/\lambda \tag{1.6}$$

Hence ν', the value of the frequency in a medium, is equal to c/λ or ν, which is the value *in vacuo*.

In describing electromagnetic radiation, therefore, the frequency is a more fundamental characteristic than is the wavelength, although in practice the wavelength measured in air is only 0·029 per cent lower than that *in vacuo*. For normal chemical purposes this error is unimportant and may be ignored.

The electromagnetic spectrum may be described in terms of a number of regions which are classified by their wavelength and frequency ranges, as shown in Table 1.1.

Table 1.1. The Electromagnetic Spectrum

Region	Wavelength range (m)	Frequency range (Hz)
Radio waves	1.0×10^0 to 1.0×10^4	3.0×10^4 to 3.0×10^8
Microwaves	1.0×10^{-3} to 1.0×10^0	3.0×10^8 to 3.0×10^{11}
Far infrared	2.5×10^{-5} to 1.0×10^{-3}	3.0×10^{11} to 1.2×10^{13}
Near infrared	7.5×10^{-7} to 2.5×10^{-5}	1.2×10^{13} to 4.0×10^{14}
Visible	4.0×10^{-7} to 7.5×10^{-7}	4.0×10^{14} to 7.5×10^{14}
Near ultraviolet	2.0×10^{-7} to 4.0×10^{-7}	7.5×10^{14} to 1.5×10^{15}
Far ultraviolet	1.0×10^{-9} to 2.0×10^{-7}	1.5×10^{15} to 3.0×10^{17}
X-ray	1.0×10^{-10} to 1.0×10^{-9}	3.0×10^{17} to 3.0×10^{18}
γ-ray	less than 1.0×10^{-10}	greater than 3.0×10^{18}

The Quantization of Electromagnetic Radiation

The divisions of the spectrum are somewhat arbitrary (except for visible light where there is little doubt about the limits), and may vary slightly with the source of the information.

Although the wavelengths in Table 1.1 are given in metres, those in the rest of the book are given in *nanometres* (1 nm = 10^{-9} m) which are more appropriate for the visible and ultraviolet regions.

The normal method of expressing frequency is in cycles per second, one cycle per second being the unit known as the Hertz (Hz). Sometimes, however, a frequency may be expressed as a *wavenumber*, which is obtained by dividing the frequency (in Hz) by the velocity of light (in cm.sec^{-1}) thus giving a result with units of reciprocal centimetres, cm^{-1}. As may be seen by rearranging equation (1.1), the wavenumber, $\bar{v}$, as given by v/c is also the reciprocal of the wavelength (expressed in centimetres). For example, visible light of a wavelength of 480 nm will have a wavenumber of

$$\bar{v} = \frac{1}{480 \times 10^{-7}} = 20{,}800 \text{ cm}^{-1}$$

and a frequency given by

$$v = \bar{v}c = 20{,}800 \times 2{\cdot}998 \times 10^{10}$$
$$= 6{\cdot}2 \times 10^{14} \text{ Hz}$$

1.2 The Nature of Electromagnetic Radiation

It must be emphasized that the description of electromagnetic radiation in terms of the wavelength or frequency of a wave must not be taken as an indication that electromagnetic radiation is propagated by a wavemotion in the same way that sound waves are. For the propagation of a wavemotion it is necessary to have a medium as, for instance, the atmosphere which carries sound waves. It is known that between the earth and the sun there is a virtual vacuum, and yet certain vital regions of the electromagnetic spectrum reach the earth from the sun without there being any detectable medium except space to support a wavemotion. This observation and the results of certain experiments, now to be described, lead to the realization that the classical wave theory of electromagnetic radiation has considerable limitations, and that many observations are inexplicable in its terms.

1.3 Energy Distribution in Black-body Radiation

A perfect radiator or 'black-body' is one in which there is equilibrium existing between radiation and matter at a given temperature. A black-body absorbs completely all frequencies incident upon it, and its emissive power for all frequencies is the maximum possible for a body at a given temperature. A hollow chamber which has a small hole in its side, maintained at a constant temperature, emits black-body radiation, the spectrum of which is independent of the material from which the chamber is constructed. The spectrum depends only upon the temperature of the enclosure. Some typical results for measurements of the intensity distribution of such radiation over various wavelengths and temperatures are shown in Figure 1.2.

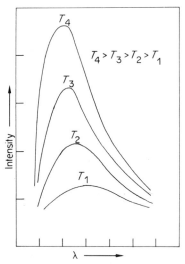

Figure 1.2. The variation with temperature of energy emitted from a 'black-body' or perfect radiator

There are three important experimentally derived pieces of information to be considered.

(1) At each temperature the intensity versus wavelength graph exhibits a maximum and, in particular, the intensity falls away as the wavelength decreases at wavelengths below that of the maximum intensity.
(2) The wavelength of maximum intensity of emitted radiation decreases with increasing temperature. It is found that the product of this wavelength, λ_{max}, and the absolute temperature, T, is a constant value. This relationship may be written in the form:

$$\lambda_{max} T = \text{constant} \quad (1.7)$$

and is known as Wien's Law.

The Quantization of Electromagnetic Radiation

(3) There is a proportionality between the total integrated intensity (area under each curve) and the fourth power of the absolute temperature, viz.:

$$I_{\text{total}} \propto T^4 \qquad (1.8)$$

which is Stefan's Law.

Using the classical wave theory, Rayleigh derived the energy distribution to have the form:

$$I_\nu \cdot d\nu = \frac{8\pi\nu^2 \cdot kT \cdot d\nu}{c^3} \qquad (1.9)$$

This equation predicts that as the square of the frequency increases (decreasing wavelength) the intensity of the emitted radiation increases,

$$I_\nu \propto \nu^2 \qquad (1.10)$$

for a given temperature. This is found to be the case at low frequencies (high wavelengths) but, of course, is not the case for higher frequencies (lower wavelengths) and the equation does not predict the maximum in the intensity versus wavelength graphs. Neither does it lead to either Wien's or Stefan's Laws.

It was in the explanation of these energy distributions that Planck introduced the *quantum theory* in which he considered radiation not to be continuous but to consist of *quanta* of *energy*. By introducing the postulate of the quantum he was able to derive an alternative equation for the energy distribution for black-body radiation. By considering the energy of such a quantum to be given by $h\nu$ the equation derived by Planck may be written as:

$$I_\nu \cdot d\nu = \frac{8\pi h \nu^3}{c^3} \cdot \left[\frac{1}{e^{h\nu/kT}-1}\right] \cdot d\nu \qquad (1.11)$$

A comparison of equations (1.9) and (1.11) reveals that if kT in Rayleigh's equation (1.9) is replaced by $h\nu/(e^{h\nu/kT}-1)$, Planck's equation (1.11) is obtained. At low values of the frequency (high wavelengths) the expression $[e^{h\nu/kT}-1]$ may be approximated to $h\nu/kT$, since in the expansion of $e^{h\nu/kT}$

$$e^{h\nu/kT} = 1 + \frac{h\nu}{kT} + \frac{(h\nu)^2}{(kT)^2 2!} + \ldots \qquad (1.12)$$

Atomic and Molecular Structure

terms above first order may be neglected for small values of $e^{h\nu/kT}$ and, in this case, Planck's expression is exactly the same as Rayleigh's, since

$$\frac{h\nu}{e^{h\nu/kT}-1} \sim kT \tag{1.13}$$

Planck's equation (1.11) predicts a maximum for the radiation intensity at a particular value of the frequency or wavelength. It also gives a sound theoretical basis to the laws of Wien and Stefan, the expressions of which, (1.7) and (1.8), are derivable mathematically from equation (1.11).

So we find in this field of study that classical theories break down, and that the experimental facts are explicable in terms of the quantum theory of radiation.

1.4 The Photoelectric Effect

The results of the investigation of the photoelectric effect are of crucial importance to our understanding of the nature of electromagnetic radiation. If radiation, say, from a mercury vapour discharge lamp, falls upon the clean surface of sodium metal (*in vacuo*), electrons are emitted from the metal surface. These are known as photoelectrons and the phenomenon of their production is known as the photoelectric effect.

The intense emission lines of a mercury vapour discharge lamp occur at wavelengths of 185, 254, 313, 365, 436, 546 and 579 nm (all quoted to the nearest 1 nm). If the radiation, emitted by the lamp, is separated into its component wavelengths by using a prism or grating monochromator and each wavelength is used in separate photoelectric experiments, it is

Table 1.2. Kinetic Energies for Photoelectrons Produced from a Sodium Surface by Mercury Radiation

Wavelength (nm)	Frequency (THz) (1 THz = 10^{12} Hz)	Kinetic energy of photoelectrons (erg $\times 10^{12}$)
597	518	zero
546	550	zero
436	688	0·870
365	822	1·764
313	958	2·669
254	1181	4·152
185	1621	7·069

The Quantization of Electromagnetic Radiation

found that no photoelectrons are produced by the radiations of wavelengths 546 and 579 nm. In these and other experiments it appears that there is a maximum wavelength above which no photoelectrons are produced. In terms of frequency this means that there is a threshold frequency below which there is no photoelectric emission.

In more sophisticated experiments the kinetic energies of the photoelectrons have been determined, and it is found that the kinetic energy is proportional to the frequency of the radiation. Such energies for the photoelectrons produced by the mercury radiation are given in Table 1.2.

A graph of the photoelectron kinetic energy versus frequency is shown in Figure 1.3 and, as can be seen from this, there is a linear relationship between the two quantities. The straight line extrapolated to zero kinetic energy intersects with the frequency axis at a value of 0.56×10^{15} Hz, which is the value of the threshold frequency for sodium as referred to above. The frequencies of the 546 and 579 nm emissions from the mercury lamp are 0.55×10^{15} and 0.518×10^{15} Hz respectively, which are both lower than the threshold frequency.

Another experimental observation is that the photoelectron energies are independent of the intensity of the radiation used. An increase in

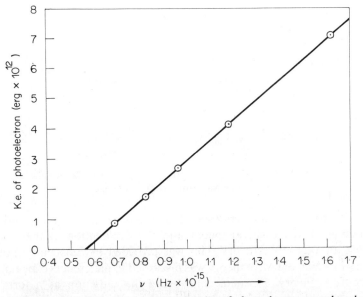

Figure 1.3. A plot of the kinetic energies of photoelectrons produced from a sodium metal surface versus the energy of the incident photons

intensity at a given frequency merely increases the number of electrons released. The experimental findings are summarized as follows:

(1) There is a threshold frequency below which no photoelectrons are produced.
(2) The photoelectron kinetic energy is directly proportional to the frequency of the radiation used and independent of the intensity.
(3) An increase in intensity causes the number of electrons released to increase.

It is interesting to compare these experimental findings with the predictions of classical wave theory. The intensity of radiation is proportional to the square of the maximum amplitude of the wave and, if photoelectron production was the result of the interaction of a wave with the metal surface, it would be expected that the kinetic energy of the photoelectrons would be proportional to the intensity of the radiation. A diagrammatical representation of this prediction is shown in Figure 1.4. No effect of frequency would be predicted and an induction period between the exposure of the metal surface to the radiation and the eventual electron release might be expected.

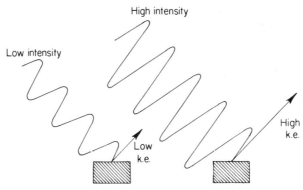

Figure 1.4. A diagrammatic representation of the 'classical' effects of two waves differing in intensity

That none of these predictions are found to be correct implies a serious breakdown of the classical theory and the explanation of the results awaited an alternative theory. This was provided by Einstein's Photon Theory in which the wave theory was discarded in preference to an almost 'particulate' theory. Einstein, considering the radiation as a form of energy, part of which is used in the breaking of the forces binding the electron to the metal surface and the excess appearing as kinetic energy

The Quantization of Electromagnetic Radiation

of the photoelectrons, noted the relationship between energy, E, and frequency, v. This relationship had been postulated previously by Planck in his explanation of the energy distribution in 'black-body' radiation. Planck's postulate may be expressed as:

$$E = hv \quad (1.14)$$

where h is the proportionality constant, known as Planck's constant.

The photoelectric effect was considered in terms of a collisional process in which a *photon* or *quantum* of energy (representing the electromagnetic energy) strikes the metal surface. A certain amount of energy, known as the *work function*, is expended in releasing the photoelectron from the surface, the excess appearing as kinetic energy of the photoelectron.

Einstein considered electromagnetic energy to be *quantized*. This means that electromagnetic energy consists of quanta or photons which are 'particulate', definite quantities of energy, the energy being determined by the frequency of the radiation as in equation (1.14). It is no longer considered to be a continuous form of energy propagated by some form of wavemotion.

The impressive simplicity of the explanation of the photoelectric effect includes the equating of the threshold frequency, v_0, with the minimum energy for electron release, i.e. the work function, W, by

$$E_{min} = W = hv_0 \quad (1.15)$$

The second step is to consider that the kinetic energy of the photoelectron will be given by the difference between the photon energy, hv, and the work function of the particular metal concerned, viz.:

$$\begin{aligned} \text{k.e.} &= hv - W \\ &= hv - hv_0 \\ &= h(v - v_0) \quad (1.16) \end{aligned}$$

This explains perfectly the graph in Figure 1.3, the slope of which gives the value of Planck's constant to be $6 \cdot 623 \times 10^{-27}$ erg.sec. This value of h corresponds exactly with the one appropriate to the treatment of black-body radiation by Planck (Section 1.3) and emphasizes the fact that h is a universal constant.

The threshold frequency, v_0, for a sodium surface is determined by the extrapolation of the graph in Figure 1.3 to zero electron kinetic energy.

These quantal ideas allow for a simple explanation of the independence of photoelectron energies upon intensity of the radiation. An increase in intensity causes more photons or quanta to strike the metal surface in unit time and has no bearing upon each individual collisional event. It would be expected from photon theory that with an increase in intensity of the radiation there would be an increase in the rate of release of photoelectrons and, indeed, this is the case. The view that each electron emitted is the result of a 'collision' between a photon and the metal surface makes the instantaneous photoelectron emission understandable.

1.5 The Compton Effect

When X-ray or γ-ray quanta interact with matter one process by which they transfer their energy to the matter is by *photoelectric absorption*. The quantum energy is transferred to an electron, the kinetic energy of such an electron being given by:

$$\text{k.e.} = h\nu - I \tag{1.17}$$

in the case of a gaseous molecule, where I is the ionization potential of the molecule concerned (this being analogous to the work function for a metal). An alternative mode of interaction between these ionizing radiations and matter is the reduction in frequency of a quantum which is observed in the Compton effect. This effect is shown diagrammatically in Figure 1.5. The incident photon has the energy, $h\nu$, and interacts with

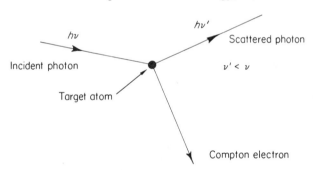

Figure 1.5. The interaction of a photon, $h\nu$, and an electron in an atom. An electron is released and the excess energy appears as another photon, $h\nu'$

a molecule in such a way as to cause ionization of an electron (the recoil electron) which possesses an amount of kinetic energy. The scattered photon has less energy than the incident photon ($h\nu$) by an amount equal to the sum of the kinetic energy of the recoil electron and the ionization potential of the molecule, i.e.

$$h\nu = h\nu' + I + \text{k.e.} \tag{1.18}$$

where I represents the ionization potential of the molecule.

The Quantization of Electromagnetic Radiation

The Compton effect is explained in terms of quantum theory in which the photons are regarded as 'packets of energy' behaving in a manner closely allied to that of particles except that their rest-mass is zero. Normal particles have masses which are related to the rest-mass by the equation:

$$m = \frac{m_0}{\left(1 + \frac{v^2}{c^2}\right)^{1/2}} \quad (1.19)$$

where m is the mass when the particle is travelling with velocity v, m_0 is the rest-mass (i.e. when $v = 0$) and c is the velocity of light. Equation (1.19) is derived from relativity theory and is important for particles travelling at high speeds.

The Compton effect is not understandable in terms of wave theory which would predict, if anything, a decrease in intensity of the wave without any alteration of its frequency.

1.6 Duality of Radiation Theories

The properties of black-body radiation, the photoelectric effect and the Compton effect have been dealt with in sufficient detail to show that the classical idea of regarding radiation as a wavemotion breaks down seriously in these cases. Some statement is required to summarize the situation of the apparently conflicting wave and photon theories. In spite of all the evidence in favour of the photon theory, there are experimental facts which can be satisfactorily explained by associating radiation with a wavemotion as, for example, interference and diffraction phenomena.

Interference phenomena may be explained by the in-phase addition of two 'waves' to give a resultant wave with a finite intensity, and by the out-of-phase addition of other waves to give a resultant wave with lower intensity than either of the two contributing waves. If the waves are 180° out of phase, the resultant wave has zero amplitude. These situations are shown diagrammatically in Figure 1.6. It is essential to grasp the true interpretation of these waves.

The wave is better regarded as not the actual form of the energy being transmitted but as a wave of probability. If electromagnetic radiation is regarded as consisting of photons, which in many ways we can think of in particulate terms, these photons behaving according to an associated probability wave, then the so-called duality of radiation theories may be satisfactorily reduced to a single viewpoint. In experiments such as the photoelectric effect and the Compton effect where the results may be

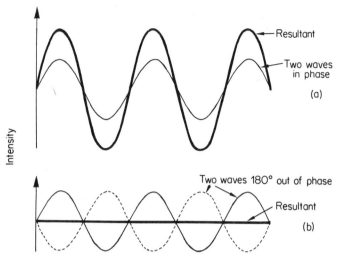

Figure 1.6. The in-phase (a) and 180° out-of-phase (b) interactions of two waves

explained in terms of single events at the atomic level, there is no need to bring in the wave part of the theory. Neither is there any need for wave arguments when dealing with photons in bulk as with the energy emitted from a perfect radiator. It is only in the more sophisticated interference and diffraction experiments (which will be dealt with in greater detail in Chapter 3) that we need to involve the enhancing and enfeebling results of the interaction of many probability waves to explain why it is more probable for the photons to end up at a particular position on the photographic plate detector than it is for other positions. In order to regard the waves as waves of probability, all we have to do is to equate probability density with the intensity of the radiation at any point at the detector, the intensity being given (as in classical theory) by the square of the amplitude of the wave.

1.7 Summary

In this chapter evidence has been presented in favour of the photon or quantum theory of electromagnetic radiation and against classical wave theory. By keeping the mathematics of the waves associated with the movement of photons, but by changing their identity from real waves to waves of probability, it has been shown that we can still think of photons when considering the phenomena of interference and diffraction. The position is one of unity and not one of dual radiation theories.

The Quantization of Electromagnetic Radiation

In the study of the properties and structures of atoms and molecules it is frequently the case that we have to regard events on the atomic scale, and it is here that quantum theory is invaluable and is entirely justified in view of the successes in explaining atomic and molecular phenomena that it achieves.

Problems

1.1 Calculate:
    (a) the frequencies,
    (b) the wave-numbers, and
    (c) the equivalent energies in ergs, electron volts, kcal.mole^{-1} and kJ.mole^{-1} of photons having the wavelengths:
       (i) 220 nm,
       (ii) 680 nm, and
       (iii) 20,000 nm.

1.2 Calculate the work function of sodium metal from the data shown in Figure 1.3.

1.3 From the data shown in Figure 1.3 check the value of Planck's constant.

2

The Quantization of Electron Energies in Atoms and Molecules

2.1 The Quantization of Electron Energy Levels

In the previous chapter experimental evidence for the quantization of electromagnetic radiation was presented and discussed in terms of the photon or quantum theory. We are now in a position to discuss the quantization of electron energy levels within the atom. This chapter will be concerned mainly with various experimental methods which all support the idea that electrons within atoms can exist only in or at certain energy levels. They cannot have a continuous range of energies but can only have certain discrete energies. This is what is meant by the quantization of electron energy levels.

Evidence for this quantization comes from the investigations in the fields of atomic emission spectroscopy, molecular absorption spectroscopy, and ionization potential data for atoms and molecules.

2.2 Hydrogen Emission Spectrum

We shall first of all consider the emission spectrum of the hydrogen atom since this is the simplest case and is understood exactly (see Section 5.3). The emission spectrum of hydrogen is produced by passing an electrical discharge through a tube in which there is hydrogen gas at low pressure (~ 1 mm.Hg). The emitted radiation may be collimated and, by using a prism or a grating, split up into its constituent frequencies, and the resulting spectrum recorded photographically. Such an emission spectrum is shown diagrammatically in Figure 2.1. A most important observation is that it consists of *lines* and a region in which there is *continuous* emission of wavelengths which cannot be resolved into lines. The discrete nature of this and other atomic emission spectra lead to the suggestion of energy levels for electrons in atoms.

The Quantization of Electron Energies in Atoms and Molecules 15

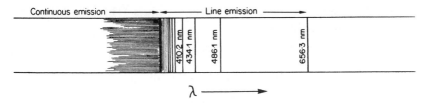

Figure 2.1. A diagrammatic representation of an atomic emission spectrum

There is a series of lines in the visible part of the hydrogen emission spectrum which have the wavelengths 656·3, 486·1, 434·1 and 410·2 nm. They form part of what is known as the Balmer series. The values of the wavelengths of the Balmer lines fit into the equation:

$$1/\lambda = R\left(\frac{1}{2^2} - \frac{1}{n^2}\right) \tag{2.1}$$

where R is a constant having a value 109666·56 cm^{-1} and n may possess the integral values 3, 4, 5, 6 .... The wavelength of the lines produced by having n equal to these various integral numbers are shown in Table 2.1.

Table 2.1. The Dependence of the Wavelength of the Balmer Lines upon n

n	Wavelength (nm)	Equivalent energy		
		(ev)	(kcal.mole^{-1})	(kJ.mole^{-1})
3	656·3	1·90	43·8	183
4	486·1	2·56	59·0	247
5	434·1	2·86	66·0	276
6	410·2	3·03	70·0	293
7	396·0	3·14	72·5	304
8	389·0	3·19	73·5	308
⋮	⋮	⋮	⋮	⋮
∞	365·0	3·40	78·4	329

As the value of n increases, it can be seen from the data in Table 2.1 that the wavelength of the corresponding line converges to a limiting value of 365·0 nm when n is infinite. The idea of energy levels in which electrons may exist within atoms is derived by converting the wavelengths of the

Balmer lines into their equivalent energy values. This is achieved by using the Planck equation:

$$E = h\nu \tag{2.2}$$

for the energy equivalent of a quantum of frequency, ν. Since

$$\nu = c/\lambda \tag{2.3}$$

then

$$E = hc/\lambda \tag{2.4}$$

and putting in values for h, c and, say, the value of λ corresponding to the first Balmer line, we have

$$E = \frac{6 \cdot 623 \times 10^{-27} \times 3 \times 10^{8}}{656 \cdot 3 \times 10^{-9}} \tag{2.5}$$

the factor of 10^{-9} being introduced so that the wavelength is expressed in metres. This equation gives a value for the energy of the quantum as:

$$E = 3 \cdot 04 \times 10^{-12} \text{ erg.atom}^{-1}$$

Using the conversion factor,

$$1 \text{ ev} = 1 \cdot 60 \times 10^{-12} \text{ erg.atom}^{-1} \tag{2.6}$$

we have

$$E = \frac{3 \cdot 04}{1 \cdot 60} = 1 \cdot 9 \text{ ev.atom}^{-1} \tag{2.7}$$

The energy corresponding to the first Balmer line is given also as:

$$E = 1 \cdot 9 \times 23 \cdot 06 = 43 \cdot 8 \text{ kcal.mole}^{-1} \quad \text{or} \quad 183 \text{ kJ.mole}^{-1}$$

The energy equivalents of the Balmer lines are shown in Table 2.1.

Bohr realized the significance of the observations. They can be interpreted as arising from electronic transitions from higher levels of energy to lower levels of energy, the difference in energy between the two levels being emitted as a quantum, $h\nu$, where

$$h\nu = E_2 - E_1 \tag{2.8}$$

E_2 and E_1 being the energies of the two levels concerned in the transition, and where $E_2 > E_1$ for emission. Equation (2.8) is known as the Bohr Frequency Condition and can be written as:

$$\Delta E = h\nu \tag{2.9}$$

The Quantization of Electron Energies in Atoms and Molecules

Applying this reasoning to the Balmer lines we realize that the 1·9 ev corresponding to the first line is the difference in energy between the two levels concerned in the particular electronic transition responsible for the emission of a quantum of wavelength 656·3 nm.

Equation (2.1) may be multiplied by the velocity of light and becomes an expression for the frequencies of the Balmer lines:

$$v = \frac{c}{\lambda} = -Rc\left(\frac{1}{n^2} - \frac{1}{2^2}\right) \tag{2.10}$$

The frequency may be expressed as the difference between the two terms, $-Rc/n^2$ and $-Rc/2^2$, the energy equivalent of the photon being the difference between the two terms, $-Rhc/n^2$ and $-Rhc/2^2$ (since $E = hv$).

Writing the equation for frequency (or energy) in this way indicates that for the Balmer lines under consideration there is a constant term, $-Rc/2^2$, and only the second (smaller) term varies. Combining this observation with the Bohr equation, (2.8), we can conclude that all the electronic transitions in the Balmer series finish in the *same* energy level, that given by

$$E = -Rhc/2^2$$

If this is the case then we can construct a diagram of energy levels with the lowest energy being given by $-Rhc/2^2$, and the successively higher energy levels being $-Rhc/3^2, -Rhc/4^2, -Rhc/5^2, \ldots$, etc., up to $-Rhc/\infty^2$ or zero which forms the reference point of the diagram shown in Figure 2.2.

The transitions responsible for the first four Balmer lines are shown in this diagram. The energy levels have been labelled with the number which in the expression for the energy level is squared, i.e. the level n, the energy of that level being given by $-Rch/n^2$. We can call n a *quantum number* and for the moment regard it as being a convenient way of labelling the electron energy levels within the hydrogen atom.

Given that the energies of the various levels are expressed by $-Rch/n^2$ where n is a quantum number, the question may be asked, 'What about $n = 1$ and the level, $-Rch/1^2$?' Surely such a beautiful explanation of the Balmer series should not have left out the energy level corresponding to the quantum number, n, having the value unity.

The latter value is indeed important and existent but does not participate in the Balmer series. It forms the lowest level in the Lyman series of hydrogen emission lines, some of whose transitions are shown in Figure 2.2.

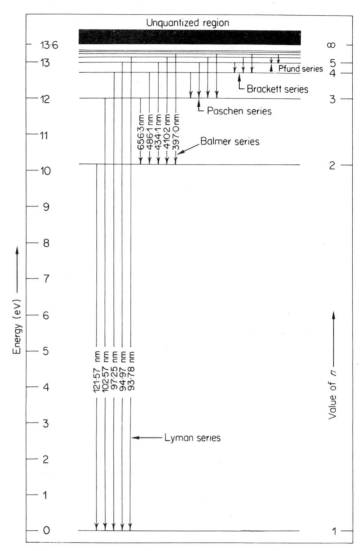

Figure 2.2. A diagram representing the line emission series of atomic hydrogen. Transitions from energies above the ionization limit ($n = \infty$) are responsible for continuous emission

The Quantization of Electron Energies in Atoms and Molecules

The frequencies of the Lyman lines are given by

$$v = \frac{c}{\lambda} = -Rc\left(\frac{1}{n^2} - \frac{1}{1^2}\right)$$

$$= Rc\left(1 - \frac{1}{n^2}\right) \tag{2.11}$$

The energy equivalents of these Lyman frequencies are given by

$$E = hv = Rch\left(1 - \frac{1}{n^2}\right) \tag{2.12}$$

where the quantum number n has values, 2, 3, 4, ..., ∞, corresponding to the wavelengths shown in Figure 2.2.

The frequency of the quantum which could cause the excitation of an electron from level, $n = 1$, to level, $n = \infty$, and which corresponds to the removal of the electron from the influence of the hydrogen nucleus is given by

$$v_{1 \to \infty} = Rc \tag{2.13}$$

This process of ionization requires an amount of energy equal to $hv_{1 \to \infty}$ or Rhc, and is calculated to be 13·6 ev or 313 kcal.mole^{-1} (1310 kJ.mole^{-1}) which agrees exactly with the experimental value.

Also shown in Figure 2.2 are three other series of emission lines where the final electronic levels are respectively $n = 3$ (Paschen), $n = 4$ (Brackett) and $n = 5$ (Pfund), and there are other series with lines occurring at even smaller energies (longer wavelengths).

In general the frequencies of the lines occurring in the emission spectrum of the hydrogen atom are given by the formula:

$$v = -Rc\left(\frac{1}{n_2^2} - \frac{1}{n_1^2}\right)$$

$$= Rc\left(\frac{1}{n_1^2} - \frac{1}{n_2^2}\right) \tag{2.14}$$

where n_1 and n_2 are different values of the quantum number, n, and where $n_1 < n_2$.

The various series of lines differ in the value of n_1, which represents the lowest energy level participating in the electronic transitions of each series. The value of n_2 is variable for each series and in all cases is greater than that of n_1.

A summary of the series of atomic emission lines for the hydrogen atom is given in Table 2.2.

Table 2.2. Line Series in the Emission Spectrum of the Hydrogen Atom

Series	n_1	n_2
Lyman	1	2, 3, 4, ..., etc.
Balmer	2	3, 4, 5, ..., etc.
Paschen	3	4, 5, 6, ..., etc.
Brackett	4	5, 6, 7, ..., etc.
Pfund	5	6, 7, 8, ..., etc.
—	6	7, 8, 9, ..., etc.
—	7	8, 9, 10, ..., etc.
	etc.	etc.

The simple diagram in Figure 2.2 explains the whole of the emission spectrum of hydrogen provided that it is considered that:

(1) the electrons may exist in discrete energy levels whose energies are governed by the value of a quantum number, n, which has integral values,
(2) in the discharge tube from which the spectrum is emitted that electrons are excited from their normal levels to upper levels and that they return to their normal levels by the transitions indicated in Figure 2.2, and
(3) the return of electrons from beyond the ionization limit ($n = \infty$) to the quantized levels give rise to the regions of continuous emission observed.

2.3 Emission Spectra of Elements other than Hydrogen

The emission spectra of the other elements may be obtained by similar methods to that which produces the hydrogen emission spectrum.

Everyone has seen the characteristic red emission of neon atoms when they are subjected to an electrical discharge and the flame test for elements is a simple application of emission spectroscopy. In a neon sign the neon atoms become electronically excited, a process which we can represent as

$$\text{Ne} \xrightarrow[\text{discharge}]{\text{electrical}} \text{Ne}^*$$

The excited neon atoms return to their normal or *ground state* by emitting a photon or quantum of energy with a characteristic wavelength. The

The Quantization of Electron Energies in Atoms and Molecules

fact that again *line* emission is observed correlates with our ideas on the hydrogen energy levels. If sodium chloride is introduced into a gas–air flame on a platinum wire, yellow light is emitted. This is characteristic of the element sodium and is a line emission. There are, in fact, two lines emitted at the wavelengths 589·0 and 589·6 nm indicating two energy levels very close together, a fact which will be discussed in detail later (see Section 6.1).

The line emission spectra of the elements serve as a very useful basis for qualitative analysis of mixtures. The spectra are highly characteristic of the elements and by comparing the spectrum of the mixture with that of a mixture producing the strong lines of many known elements a rapid qualitative analysis of the unknown mixture can be made.

For our purposes, however, it is only necessary to record that all elements give *line* emission and that the phenomena seems to be universal. We can conclude that there is a tremendous amount of evidence from this branch of spectroscopy which supports the theory that electrons exist in energy levels.

The mathematical relationships which summarize the various series of lines in the spectrum of a particular element are similar to the formulae which serve for hydrogen but are not quite as simple, the n values (of equation 2.14) no longer being integral. In such cases the actual quantum numbers (n) are modified by a non-integral amount to reproduce the frequencies observed in any particular series. Deviations from the very simple behaviour of the hydrogen spectrum always occur when the atom in question contains more than one electron. Electron–electron interaction which is of great importance in chemistry is dealt with at length in Section 6.2.

2.4 Molecular Absorption Spectra

In the Bohr expression for emission of a quantum of radiation, equation (2.8), E_2 was the energy of the initial state and was higher than E_1, the energy of the final state, so that in the transition, $2 \to 1$, energy was released in the form of a quantum. Exactly the same expression holds for absorption of a quantum; hv causing an electron to be excited from $1 \to 2$ with exactly the same change in energy.

The Bohr equation is the expression of the condition for absorption by atoms or molecules of quanta. Absorption will only take place when there exists an energy level which is higher than the one in which the electron to be transferred resides by an amount equivalent to the energy of the quantum used in the experiment. Quanta with energies lower than or higher than this specific energy difference will not be absorbed. If, in fact, this is a true representation of the situation in atoms and molecules, then the absorption of radiation by such species should show a variation with wavelength of the radiation and give rise to *absorption spectra*.

Three examples will suffice to demonstrate this point. The first is the absorption spectrum of benzene vapour as shown in Figure 2.3, where the *absorbance* is plotted against the wavelength. The Beer–Lambert Law of light absorption may be expressed as:

$$I_T = I_0 10^{-\varepsilon c l} \tag{2.15}$$

where I_T is the intensity of transmitted radiation of wavelength λ, I_0 is the intensity of incident radiation of wavelength λ, ε is the molar absorption coefficient ($l.\text{mole}^{-1}.\text{cm}^{-1}$), c is the concentration of absorbing species in $\text{mole}.l^{-1}$, and l is the length of the absorbing path in cm. By taking

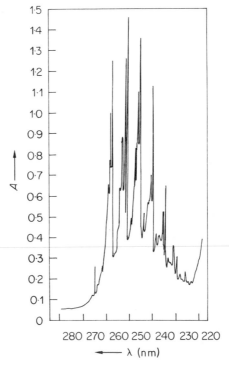

Figure 2.3. The longest wavelength band in the electronic spectrum of benzene vapour

The Quantization of Electron Energies in Atoms and Molecules

logarithms of both sides of equation (2.15), we obtain

$$\log_{10} I_T = \log_{10} I_0 - \varepsilon cl \tag{2.16}$$

which may be rearranged to:

$$\log_{10} I_0/I_T = \varepsilon cl \tag{2.17}$$

The quantity $\log_{10} I_0/I_T$ may be defined as the *absorbance*, A, of the absorbing material in the cell, so that

$$A = \varepsilon cl \tag{2.18}$$

The absorbance at any fixed wavelength is proportional to the molar concentration, c, of the absorbing substance and the optical path length, l, of the cell containing the substance. This form (equation 2.18) of the Beer–Lambert Law is of inestimable use in analytical chemistry.

As can be seen from Figure 2.3, the spectrum of benzene vapour is a spiky spectrum again fitting in with the theory of discrete electron energy levels. In fact, this absorption shown is one electronic transition, and the reason for its not being one line as with atomic spectra is that in the transition the vibrational and rotational energy of the molecule may change in a variety of ways.

The vibrational energy of a molecule is quantized and the equation,

$$E_{\text{vib}} = (v + \tfrac{1}{2})h\omega_0 \tag{2.19}$$

represents the allowed values for a diatomic molecule with a fundamental vibration frequency ω_0. The vibrational quantum number may have the values, $0, 1, 2, 3, \ldots$, and it is important to note that even when $v = 0$, $E_{\text{vib}} = \tfrac{1}{2}h\omega_0$, i.e. there is *zero-point* vibrational energy even when the molecule is in its lowest state of energy.

The rotational energy of a diatomic molecule is also quantized and is given by the equation,

$$E_{\text{rot}} = \frac{J(J+1)h^2}{8\pi^2 I} \tag{2.20}$$

where J is the rotational quantum number which may have the values, 0, 1, 2, etc., and I is the moment of inertia of the molecule and is equal to μr^2 where μ is the reduced mass of the molecule and r is the internuclear distance (bond length).

The total energy of a molecule is made up from electronic, vibrational and rotational contributions:

$$E_{\text{total}} = E_{\text{elec}} + E_{\text{vib}} + E_{\text{rot}} \tag{2.21}$$

and in an electronic transition all three of these contributions are changed, resulting in complex spectra even in the case of a diatomic molecule.

In the case of benzene (Figure 2.3) what is observed is the lowest energy electronic transition, superimposed on which are several vibrational–rotational bands. The resolving power of the instrument used to record this spectrum is not sufficiently high to show the details of the rotational fine structure of each vibrational band, which therefore appear as wide bands and not lines.

The second example is the absorption spectrum of formaldehyde. There is an absorption band with a maximum absorption at 280 nm and another with a maximum absorption at 170 nm. The two transitions responsible for these absorptions are:

(1) the transfer of a lone pair (non-bonding) electron on the oxygen atom to an upper level (π^*); and
(2) the transfer of an electron concerned in the bonding of the carbon and oxygen atoms to the same upper level.

These transitions are shown diagrammatically in Figure 2.4. Energy calculations using the λ_{max} shows that the equivalent energies of quanta

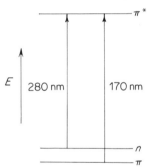

Figure 2.4. The $n \to \pi^*$ and $\pi \to \pi^*$ transitions in the electronic absorption spectrum of formaldehyde

of wavelengths of 280 nm and 170 nm are 4·44 ev and 7·32 ev respectively, indicating the difference in energy between the non-bonding and bonding electronic levels to be $7·32 - 4·44 = 2·88$ ev.

The levels in Figure 2.4 are labelled π^*, n and π in order of decreasing energy. Such terms will be fully explained in Section 7.4.

It would appear then that observations of molecular absorption spectroscopy produce additional evidence of the existence of energy levels

The Quantization of Electron Energies in Atoms and Molecules 25

which may be occupied when the molecules are in an electronically excited state and also for energy levels which are occupied in the ground state of the molecule.

In the case of gaseous molecules, the absorption spectrum consists of a line spectrum in which the frequencies of the series of lines may be assigned to the formula,

$$v = A - \frac{B}{(n+m)^2} \qquad (2.22)$$

A and m being constants for a given series, and B being a universal constant. The constant m is usually negative and represents the deviation of the sum of n and m from integral value: it is known as the quantum defect. When n, the quantum number, becomes large the lines converge to a limit, v_∞, where $n = \infty$ and $v_\infty = A$. Such a series of lines whose frequencies converge to the ionization limit (v_∞) is known as a Rydberg series. Other examples which have been mentioned are the Lyman, Balmer and Paschen series of the hydrogen atom (usually seen by emission). The ionization potential of the molecule will be given by hv_∞. Again, evidence for discrete energy levels for electrons in excited states is obtained as for atoms. Other series lead to different values for the ionization potential and lead to the conclusion that a molecule may have more than one ionization potential—in fact as many ionization potentials as there are energy levels. This is shown by the diagram in Figure 2.5 for a molecule with four energy levels, in which an electron may be ionized in

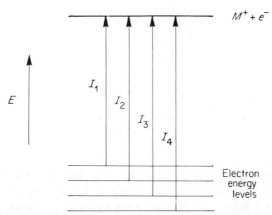

Figure 2.5. The four ionization potentials of a molecular system in which the electrons are accommodated in four levels of different energy

turn from the levels, *A*, *B*, *C* and *D*, by using larger and larger amounts of energy.

2.5 Ionization Potential Data for Atoms

Important information concerning the distribution of electrons within atoms can be gained from ionization potential measurements. The ionization potential of an atom may be defined as the energy required to completely remove an electron from the ground state of the free atom in its gaseous state.

A mass spectrometer may be used to:

(1) produce the ions by bombarding the atoms with electrons of known kinetic energy, and
(2) detect ionic species with different m/e ratios. It is possible to record the ion-current produced by a given electron energy, the result being shown in Figure 2.6.

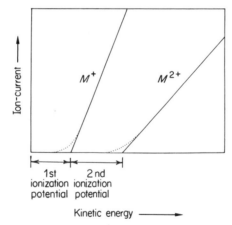

Figure 2.6. Plots of ion-current for the ions M^+ and M^{2+} versus the energy of the bombarding electrons

The ion-current versus electron energy plots are usually linear except for a curved region near what may be called the appearance potential. Extrapolation of the linear part of the plot to zero ion-current gives a measure of the minimum energy required to cause ionization of the gaseous species, i.e. the ionization potential.

The method suffers from being a threshold technique in that the ionization potential is the minimum energy that will cause ionization. In fact

The Quantization of Electron Energies in Atoms and Molecules 27

there is, for a given kinetic energy of the electrons, a slight variation (± 0.1 ev) in this energy which causes the curvature shown in Figure 2.6. The energy required to ionize the atom of argon,

$$Ar_{(g)} \rightarrow Ar^+_{(g)} + e^- \tag{2.23}$$

to produce a unipositive argon ion is found mass spectrometrically to be 15·775 ev, and is known as the first ionization potential since it is the amount of energy concerned in removing the first and least stable electron from the argon atom. The second ionization potential is the amount of energy required to carry out the process,

$$Ar^+_{(g)} \rightarrow Ar^{2+}_{(g)} + e^- \tag{2.24}$$

and is found to have the value 27·62 ev. This does not necessarily mean that the second easiest electron to remove from argon comes from a level of energy which is below that of the first electron, since in process (2.24) a negative electron is being removed from an already positively charged ion, whereas in process (2.23) the removal is from a neutral atom. The added electrostatic attraction of the ion for the electron which is removed is the cause of the increase between the first and second ionization potentials, and is a general effect when successive ionization potentials are considered. As the ionization steps proceed, the effectiveness of the nuclear charge increases and causes a general increase in the successive ionization potentials.

Let us now look at the successive ionization potentials of the atom of potassium as shown as a plot of the logarithm of the ionization potential against the number of electrons removed in Figure 2.7. The logarithmic plot is used for convenience only since the range of values (5–5000 ev) is considerable. There are three large increases superimposed on the expected increase as the number of successive electrons removed increases. It would seem that the first electron is removed with comparative ease (4·4 ev) while the second electron requires 31·8 ev for its removal. The next seven electrons are removed with increasing difficulty but there are no glaring discontinuities in the graph until the tenth electron is removed. The average increase in adjacent ionization potentials of the electrons $2 \rightarrow 9$ is 22·6 ev, whereas the gap between the ninth and tenth potentials is 280 ev, corresponding to a greater increase in the effectiveness of the nuclear charge. The eleventh to seventeenth electrons are removed with a regularly increasing difficulty, and then there is an increase of about 4000 ev to the eighteenth potential compared with an average increase of 74 ev over the eleventh to seventeenth, indicating the operation of another extraordinary increase in the effective nuclear charge.

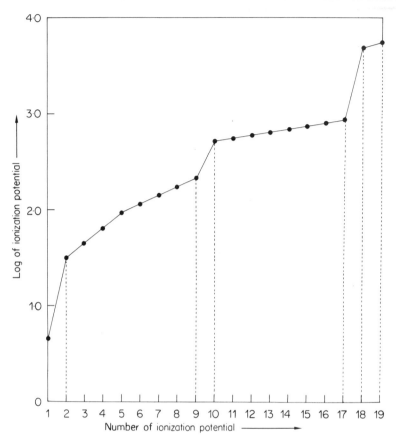

Figure 2.7. A plot of the logarithm of the successive ionization potentials of the potassium atom against the number of electrons removed. (From page 111 of the *Journal of the Royal Institute of Chemistry*, 1963; reproduced by permission of the Royal Institute of Chemistry)

Interpreting the discontinuities as indicating the points at which one electron level is vacant and the next electron comes from a lower, more stable, energy level, it is possible to draw an energy level diagram such as the one in Figure 2.8 in which the nineteen electrons of potassium are distributed in the four energy levels as shown.

Two electrons are placed in the lowest, most stable, level; these are the last to be removed. In the next highest level there are eight electrons followed by another eight in a level even higher in energy. Finally the single 'valency' electron resides in the highest occupied level. This can be considered as being further away from the nucleus whose effective

The Quantization of Electron Energies in Atoms and Molecules

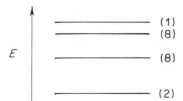

Figure 2.8. A possible distribution of electrons in the potassium atom

charge will be reduced by the intervening eighteen electrons and will be the easiest to ionize.

The second electron to be removed would be more difficult because of the increased effectiveness of the nuclear charge which is considerable because the electron to be removed in the second ionization process exists in a lower energy level than did the first electron. The next seven

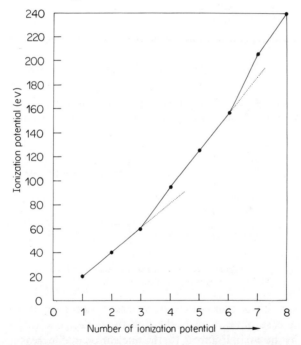

Figure 2.9. The first eight successive ionization potentials of the neon atom plotted against the number of electrons removed. (From page 111 of the *Journal of the Royal Institute of Chemistry*, 1963; reproduced by permission of the Royal Institute of Chemistry)

potentials would increase regularly, but the tenth electron would come from the next more stable energy level (about 250 ev lower than the previous level) and would be more difficult to remove.

Following a regular increase in the effectiveness of the nuclear charge in the removal of the eleventh to the seventeenth electrons there would be a more than average increase in the ionization potential of the eighteenth electron since it would have to be removed from the very stable doubly occupied level.

Accurate measurements of the successive ionization potentials of a lighter element such as neon can give more information concerning the groups of eight, and the values for the first eight successive ionization potentials of neon are shown in Figure 2.9. The detailed consideration of the group of eight electrons shows that there are two discontinuities in the general increase in the values of the successive ionization potentials. It would appear that we should split the group of eight electrons into sub-levels. Also we may deduce that the electrons exist in pairs in close proximity, and whenever there are two electrons in such a situation there will be a certain amount of electrostatic repulsion between them, causing the removal of the first electron to be slightly easier than the removal of the second, which no longer has the electrostatic assistance in its progress from the particular energy level. For neon, then, we can place the eight outermost electrons in levels, as represented in Figure 2.10, as two sub-levels, the lower one containing two electrons and the upper

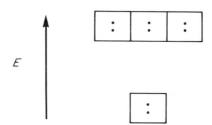

Figure 2.10. A possible electron distribution for the neon atom

one containing three pairs of electrons. Electrostatic repulsive assistance explains the ease of removal of the first three electrons to give the configuration as shown in Figure 2.11, the nuclear charge increasing regularly in its effectiveness.

The next three electrons would not have the advantage of electrostatic assistance and are more difficult to remove. This explains the slight discontinuity between the third and fourth ionization potentials. The

The Quantization of Electron Energies in Atoms and Molecules

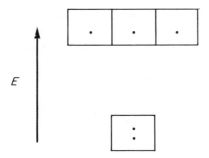

Figure 2.11. The electron distribution in the Ne^{3+} ion

second greater discontinuity comes between the sixth and seventh ionization potentials and is due to the seventh and eighth electrons being in a lower sub-level than the first six.

The arrangement of eight electrons as shown in Figure 2.10 can explain the variation in the first ionization potentials of the elements of the first short period of the periodic table, lithium to neon. The data are plotted in Figure 2.12. Starting with lithium the first electron will reside in the

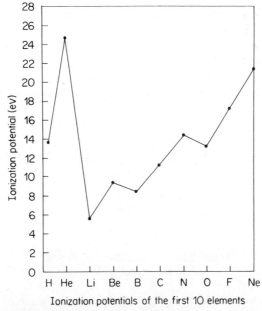

Ionization potentials of the first 10 elements

Figure 2.12. The first ionization potentials of the ten atoms, hydrogen to neon

lower sub-level in the ground state (two other electrons exist in a very much more stable level with respective ionization potentials of 75·6 and 122·4 ev). In beryllium the same sub-level will contain two electrons, the first ionization potential being almost twice that of the lithium atom. The reason for this increase is the considerable increase in nuclear charge from +3 to +4 in going from lithium to beryllium, which is more than enough to offset any electrostatic repulsive assistance offered to the electron leaving the beryllium atom.

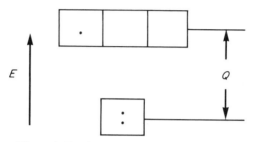

Figure 2.13. The electron distribution of the boron atom

The configuration of the boron atom would be as in Figure 2.13 and in spite of the increase in nuclear charge to +5, the first ionization potential is lower than that of beryllium because of the energy gap, Q, between the two sub-levels concerned. The first ionization potentials of carbon and nitrogen show increases because of the increasing nuclear charge. With oxygen we have the arrangement shown in Figure 2.14.

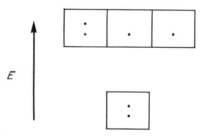

Figure 2.14. The electron distribution of the oxygen atom

Two electrons are paired up in one 'compartment' whereas in nitrogen the three electrons in the upper sub-level could minimize the repulsion between them by singly occupying all three 'compartments'. It is with

The Quantization of Electron Energies in Atoms and Molecules

oxygen that the electrostatic repulsive assistance overcomes the increase in nuclear charge ($+7 \to +8$) and causes the first ionization potential of oxygen to be less than that of nitrogen. In fluorine and neon there are regular increases due to the nuclear charge increases.

Ionization potential measurements provide:

(1) evidence for the existence of electrons in discrete energy levels within ground state atoms, and
(2) evidence for the existence of electrons in 'compartments' which may contain one or, at most, two electrons, with the configurations involving unpaired electrons being preferred on energetic grounds.

2.6 Ionization Potential Data for Molecules

Mass spectrometry may be used to determine the ionization potentials of molecules as it is for those of atoms (Section 2.5). The determination of the first ionization potential of the nitrogen molecule is shown in Figure 2.15 where the ion current produced by N_2^+ ions is plotted against the energy of the bombarding electrons.

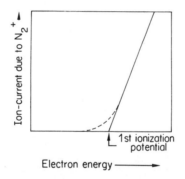

Figure 2.15. A plot of the ion-current due to N_2^+ ions versus electron bombarding energy— measurement of the first ionization potential of N_2

Another method for the determination of ionization potentials is to use a photoionization technique. In this a minimum wavelength is found which will cause ionization of the molecule. The energy equivalent of this wavelength is then taken to be the ionization potential of the molecule.

Both these methods are 'threshold' techniques and are subject to slight inaccuracies as a result of this. A recent and non-threshold technique for the determination of molecular ionization potentials is that known as molecular photoelectron spectroscopy. In principle the method used is to cause ionization of a gaseous molecule with the 56·4 nm line from a helium discharge lamp. The spectrum of photoelectron energies is measured and for the hydrogen molecule, for example, there is a maximum number of electrons with an energy of 5·8 ev. Since the ionizing radiation has an energy equivalent to 21·21 ev the first ionization potential of the hydrogen molecule is equal to 21·21 − 5·8 = 15·41 ev, which agrees closely with the value of 15·422 ev derived from the convergence limit of the first Rydberg series of the molecule.

With more complex molecules it is possible for this method to give accurate values for more than one ionization potential, as in the case of nitrogen where the values of 15·57, 16·72 and 18·72 ev are found. Other examples are shown in Table 2.3.

Taking the ionized state as the reference zero, it follows that these discrete ionization potentials may be used to draw diagrams of energy levels for electrons within molecules, as shown in Figure 2.16 for oxygen, nitric oxide and nitrogen.

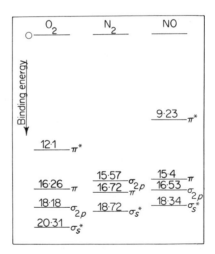

Figure 2.16. Some of the electron energy levels for the molecules O_2, NO and N_2 as measured by photoelectron spectroscopy

The Quantization of Electron Energies in Atoms and Molecules

Table 2.3. The Ionization Potentials of Some Small Molecules Given by Photoelectron Spectroscopy

Molecule	Ionization potentials (ev)
C_2H_2	11·36, 16·27, 18·33
CO	13·98, 16·58, 19·67
NO	9·23, 15·4, 16·53, 18·34
O_2	12·1, 16·26, 18·18, 20·31
H_2O	12·61, 14·23, 18·02
H_2S	10·42, 12·62, 14·82, 20·12

From this diagram it is obvious that the highest electron energies in NO and O_2 are considerably higher than those in nitrogen. These observations are consistent with the observed greater chemical reactivities of oxygen and nitric oxide compared to the nitrogen molecule. Even the highest energy electrons in the nitrogen molecule are at low energy compared to those in the oxygen and nitric oxide molecules. A detailed discussion of the energy levels involved is to be found in Section 7.8.

Problems

2.1 Plot the first ionization potentials of the first thirty-six elements against the nuclear charge. Notice the periodicity in the main group elements and the different pattern for the transition elements (data obtainable from Figure 6.10).

2.2 Plot the successive ionization potentials of the elements, Li, Be, B, C, N, O and F, and show how the variations can be interpreted in terms of electronic energy levels (data obtainable from Appendix II).

2.3 Plot the ionization potentials of He^+, Li^{2+}, Be^{3+}, B^{4+}, C^{5+}, N^{6+}, O^{7+} and F^{8+} against the square of the nuclear charge. Determine the slope of the graph and hence derive a value for the Rydberg Constant.

2.4 Plot the ionization potentials of Li^+, Be^{2+}, B^{3+}, C^{4+}, N^{5+}, O^{6+} and F^{7+} against the square of the nuclear charge. Determine the slope of the graph and compare its value with that of the Rydberg Constant. Discuss the reason for the deviation and calculate the amount that Z (in each case) has to be modified by so that the equation $I = R(Z')^2$ is satisfied, with R having its normal value of 13·6 ev. Compare the values of Z' with those of Z and suggest an interpretation of the differences.

2.5 The hexacyanoiron (III) ion, $Fe(CN)_6^{3-}$, possesses a molar absorption coefficient of $1·3 \times 10^3$ $l.mole^{-1}\,cm^{-1}$ at 313 nm. If a solution containing this ion has an absorbance of 0·74 in a 5 cm cell, calculate the concentration of the ion in the solution. Assume that no other species present absorb at 313 nm. Calculate the percentage transmission ($100\,I_T/I_0$) of a $3·85 \times 10^{-3}$ M solution of the ion in a 0·1 cm cell at 313 nm.

2.6 A solution of potassium chromate ($5·0 \times 10^{-5}$ M) in a 4 cm cell has an absorbance of 0·6 at 365 nm. Calculate the molar absorption coefficient of this compound.

2.7 The fundamental vibration frequency of the hydrogen chloride molecule is $8·66 \times 10^{13}$ Hz. Using equation (2.19) calculate the energy required to cause the $v = 0$ to $v = 1$ transition.

2.8 The $J = 0$ and $J = 1$ transition in the carbon monoxide molecule requires an energy of $10·88$ $cal.mole^{-1}$. Using equation (2.20) calculate:
(a) the moment of inertia, and
(b) the bond length of the molecule.

2.9 Compare the energy of the electronic transition which occurs when the hexacyanoiron (III) ion absorbs radiation of wavelength 313 nm (problem 2.5), with those of the vibrational transition of the HCl molecule (problem 2.7), and of the rotational transition of the carbon monoxide molecule (problem 2.8).

3

The Diffraction of Electrons and the Uncertainty Principle

3.1 Electron Diffraction

In Chapter 1 we saw that classical ideas on electromagnetic radiation had to be revised in view of the results of experiments in various fields of study. The new quantum or photon theory was based on a more particulate view of radiation and the wave behaviour was interpreted in terms of the probability of finding photons at any given point.

It is normal to consider electrons as particles. They may be shown to possess mass. If an electron strikes a fluorescent screen, which may be made from zinc sulphide, a scintillation or flash of light is observed at the point of collision. The effect of the collision appears to be due to a discrete particle without any widespread interaction with the zinc sulphide screen.

However, it was observed by Davisson and Germer in 1927 that, if a beam of electrons of known kinetic energy were allowed to strike the face of a crystal of nickel metal, the reflected electron pattern as detected on a photographic plate strongly resembled the pattern obtained by diffracting a beam of monochromatic radiation. This experiment is shown diagrammatically in Figure 3.1.

Another experiment involving electron diffraction was carried out in 1928 by Thomson and Reid and consisted of the bombardment of a thin gold foil screen by an electron beam, the photographic plate being exposed at the other side of the foil. Again an electron diffraction pattern was obtained.

An example of such a pattern is shown in Figure 3.2 which is the pattern obtained by passing electrons of an energy of 36,000 ev through thin silver foil.

We are now faced with the problem of explaining how 'particles' can produce a diffraction pattern—something that we link with wave

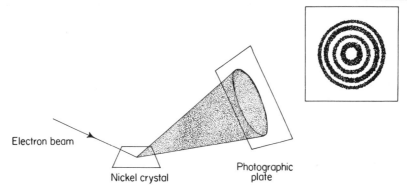

Figure 3.1. A diagrammatic representation of the diffraction of an electron beam by a metal surface

Figure 3.2. An example of an electron diffraction pattern. This was produced by passing 36,000 ev electrons through silver foil. (From *Atomic Spectra and Atomic Structure*, by Gerhard Herzberg, Dover Publications, Inc., New York, 1944; reprinted through permission of the publisher)

behaviour. If, in the experiments quoted above, the photographic detector is replaced by a scintillation screen, it can be demonstrated that the electrons after diffraction still produce point scintillations. They are still discrete amounts of matter. If the scintillation pattern is put on record and these records for a finite diffraction time are superimposed, the result in terms of scintillation per unit area is identical to the density per unit area of the photographic plate detector.

We can conclude from this that although electrons are particles of matter, they behave under certain circumstances as though their motion was governed by a wave. From the diffraction pattern in Figure 3.2, it may be calculated that the wave associated with the 36,000 ev electrons has a wavelength of 0·00645 nm.

3.2 Matter Waves

In 1924, before the observation of the 'wave' properties of the electron, de Broglie had postulated that whenever a particle of mass, m, moved with a velocity, v, there would be an associated *matter wave*.

The argument of de Broglie began with the Planck equation,

$$E = h\nu \tag{3.1}$$

relating energy with frequency of radiation. Einstein's Theory of Relativity provides the equation,

$$E = mc^2 \tag{3.2}$$

which expresses the energy equivalent to an amount of matter with mass, m, c being the velocity of electromagnetic radiation.

The right-hand sides of equations (3.1) and (3.2) may be equated to give equation (3.3):

$$h\nu = mc^2 \tag{3.3}$$

For photons we may use the equation,

$$\lambda = c/\nu \tag{3.4}$$

to express the wavelength in terms of the frequency of the radiation and its velocity. If we now use the value of the frequency from equation (3.3) in equation (3.4), we derive equation (3.5):

$$\lambda = h/mc \tag{3.5}$$

So far the equations used have been theoretically exact and the derived relationship, equation (3.5), must also be exact, but only for photons. The important step taken by de Broglie was to replace c by v in equation (3.5) so that

$$\lambda = h/mv \tag{3.6}$$

applied to particles of mass, m, moving with a velocity, v, where $v < c$, as must be the case for all particles except photons.

Equation (3.6) may be rewritten as:

$$\lambda = h/p \tag{3.7}$$

where p is the momentum of the moving particle ($p = mv$).

The justification for equation (3.7) came when it was found that the wavelengths of the electrons derived by experiments similar to those already described in Section 3.1 were predicted accurately.

The kinetic energy of the electrons in the electron diffraction experiment described above (Section 3.1) was known to be 36,000 volts. We may write this in ergs as:

$$\text{k.e.} = \tfrac{1}{2}mv^2 = 36{,}000 \times 1{\cdot}6 \times 10^{-12}\ \text{erg} \tag{3.8}$$

The momentum, mv, of the electrons is given by

$$mv = (\tfrac{1}{2}mv^2 \times 2m)^{1/2} = (\text{k.e.} \times 2m)^{1/2}$$

$$= (36{,}000 \times 1{\cdot}6 \times 10^{-12} \times 2 \times 9{\cdot}1 \times 10^{-28})^{1/2}$$

$$= 1{\cdot}02 \times 10^{-17}\ \text{g.cm.sec}^{-1} \tag{3.9}$$

Using the de Broglie relationship (equation 3.6) we can now calculate the equivalent wavelength of the electron beam to be

$$\lambda = \frac{h}{mv} = \frac{6{\cdot}623 \times 10^{-27}}{1{\cdot}02 \times 10^{-17}}$$

$$= 6{\cdot}45 \times 10^{-10}\ \text{cm}$$

$$= 0{\cdot}00645\ \text{nm} \tag{3.10}$$

The agreement between experiment and theory thoroughly justifies de Broglie's equation.

There is no doubt that the behaviour of electrons in diffraction experiments may be described by a waveform with a wavelength given by equation (3.7). The danger lies in attributing wave properties to electrons. Electrons are best described as particles. They are not waves. Their motion may be represented by wave equations which are to be interpreted as being methods of calculating the probabilities of finding electrons at, say, points on the photographic plate or the scintillation screen.

The diffraction pattern produced by a beam of electrons passing through a thin metal foil may not be interpreted in a statistical sense as being produced by the interaction of a vast number of electrons. The pattern produced is independent of the intensity of the electron beam and would be the same (for the same number of electrons) even if the electrons were diffracted one at a time. The 'wave' properties of electrons are not due to bulk effects but are a built-in property of each individual electron.

The behaviour of electrons in atoms may be represented by a so-called wave equation. It should be understood that this does not in any way imply wave behaviour of the electrons, but should be interpreted as a method of calculating electron probabilities. There are analogies between electron 'waves' and those which are used to describe electromagnetic

The Diffraction of Electrons and the Uncertainty Principle

radiation. The electrons do not undergo any form of oscillation. Any oscillation which occurs is the mathematical undulation of the function which is concerned with the calculation of the probability of finding an electron with given values of the coordinates of the system. The derivation of such functions and the necessity for so doing is dealt with in Chapter 4.

3.3 The Observation of Electrons and the Uncertainty Principle

If electrons are detected by using a scintillation screen they appear to be particulate since for each collision involving one electron a single flash of light is produced. The effect is localized at the point of contact with the screen. If, however, as in the diffraction experiments already described, a photographic plate detector is used, a diffraction pattern is obtained which leads to the wave ideas for electron motion.

Similar dual behaviour is observed with electromagnetic radiation, wave properties being inferred as a result of diffraction experiments and the 'particulate' properties being apparent in the photoelectric experiment.

It would appear, therefore, that since various methods of observing electrons (or photons) lead to either wave or particulate views of their nature, the actual event of observation causes this variation of interpretation of the results since one cannot imagine that electrons or photons change their properties according to the method used to observe them. The situation is that part of the nature of electrons is revealed by one method of observation and another part by another method, neither method revealing the true nature of the electron (or photon).

The majority of methods of observation involve the use of photons which are used to illuminate the object. After bouncing off the object the photons are transmitted to a recording device which may be a photographic plate, a photoelectric cell or the eye. Such methods are satisfactory for the majority of objects. The photons which are used do not in any measurable way affect the object. The situation is different with microscopic objects such as atoms and electrons. Electrons cannot be 'seen' in the normal sense of that word. They are too small. To observe such small objects we should have to use a microscope with a resolving power greater than any actual microscope now in existence.

A useful exercise is to carry out a *'thought experiment'* in which an ideal (non-existent) microscope is used to view an electron. In order that the resolving power be suitable we have to use electromagnetic radiation with a wavelength equal to or smaller than the object to be observed. To maximize the resolving power of our ideal microscope we may think of using short wavelength gamma rays. These are photons with extremely large energies, and if they are used to strike the electron they will interact in such a way as to alter the momentum of the electron in some imprecise way. The momentum of the electron will be altered to an indeterminate extent. In this experiment we may have observed the electron but the process of 'seeing' has altered the electron's momentum, and we come to the conclusion that if the *position* of the electron is known, the momentum is uncertain.

Next consider the measurement of the velocity of the electron. To do this we have to observe the electron twice in timing its motion through a given distance. As we have already concluded the process of 'seeing' involves the transference of indeterminate amounts of energy to the electron and thus alters its momentum. To minimize this we can use extremely long wavelength, low energy photons in the ideal microscope. This would ensure that the uncertainty in the momentum was minimal but, of course, with such long wavelengths having to be used, the resolving power of the microscope is reduced to a minimum and it would not be possible even with our 'ideal' system to observe the position of the electron with any certainty. We conclude from this that if the momentum of the electron is known accurately, it is not possible to know its position with any certainty.

The two conclusions we have reached are summarized as, and follow from, Heisenberg's *Uncertainty Principle* or *Principle of Indeterminacy* which may be stated in the form: 'It is impossible to determine simultaneously the position and momentum of an atomic particle.'

A very straightforward mathematical derivation of the principle can be carried out by considering the phenomena of the diffraction of an electron. When an electron beam is directed towards a thin foil as in Figure 3.3, diffraction occurs and electrons are detected on the photographic plate having been diffracted away from their normal path. This may be interpreted in terms of a particular electron receiving extra momentum equivalent to Δp where

$$\Delta p \sim p \sin \alpha \qquad (3.11)$$

The minimum value of Δp will be zero which corresponds to the normal (undiffracted) path of the electron. Δp represents the uncertainty in the momentum of the electron at the time of its diffraction through the gap which in size is Δq. The latter quantity is an expression of the uncertainty in the position of the electron at the time of its diffraction. From diffraction theory the angle, α, is of the order of $\lambda/\Delta q$, so that for small slits and long wavelengths the diffraction angle is greater. Providing α is small, we can write:

The Diffraction of Electrons and the Uncertainty Principle

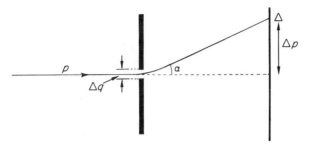

Figure 3.3. A diagram showing the path of an electron, which is diffracted by passing through a narrow opening. The size of the opening represents the uncertainty in the position of the electron, Δq, at the time of diffraction

$$\sin \alpha \sim \alpha \tag{3.12}$$

and

$$\Delta p \sim p\alpha$$
$$\sim p\lambda/\Delta q \tag{3.13}$$

or

$$\Delta p \cdot \Delta q \sim p\lambda \tag{3.14}$$

Using the de Broglie equation,

$$p = h/\lambda \tag{3.15}$$

in equation (3.14), we get

$$\Delta p \cdot \Delta q \sim h \tag{3.16}$$

which is a simple mathematical expression of the uncertainty principle. Δp represents the uncertainty in the momentum of the electron and Δq the uncertainty in its position, the product of these uncertainties being of the order of Planck's constant, $6 \cdot 624 \times 10^{-27}$ erg.sec.

In terms of a probability wave, to have exact knowledge of the momentum it is necessary to know the wavelength, λ. This means that we must describe the electron by a simple sine wave of known wavelength. With the principle that the squared amplitude of this wave represents the probability of finding the electron, we conclude that wherever there is a maximum or a minimum in the wave equation, there is an equal chance of the electron being in that position. In Figure 3.4, for instance, the wave equation is represented by curve A, the square of this being

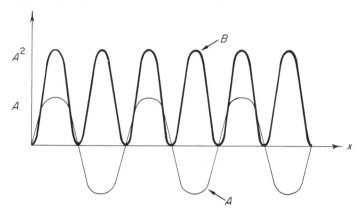

Figure 3.4. A monochromatic sine wave (curve *A*) together with its square (curve *B*)

shown as curve *B*. At the maxima in the 'squared' wave (curve *B*) there would be equal and maximal probabilities of finding the electron. Since a sine wave extends from minus infinity to plus infinity there is therefore an infinite number of the maxima shown in curve *B*. In other words, there is complete uncertainty in the *position* of the electron as described by a monochromatic sine wave whose definite wavelength, λ, determines the momentum of the electron exactly.

Taking next the case where the position of the electron is known exactly we must deal with a wave such as the one in Figure 3.5 where

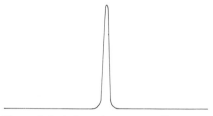

Figure 3.5. A 'wave' corresponding to an electron whose position is known exactly

there is one very narrow peak indicating accurately the position of the electron. Such a wave may be represented as being the resultant of the interaction of many sine waves with as many different wavelengths. Such knowledge can be derived by a Fourier analysis of the equation describing the 'wave' in Figure 3.5. So, although the position is known with accuracy,

The Diffraction of Electrons and the Uncertainty Principle

there is complete uncertainty in the momentum since the 'wave' is a composite one of waves of many different wavelengths.

3.4 Consequences of the Uncertainty Principle

The principle of indeterminacy has far reaching consequences in our dealings with atoms and molecules. The main consequence is that models of the atom in which electrons are assigned to orbits with a specific radius (i.e. of exactly known position) are not valid within the terms of the uncertainty principle. The main consequence of the operation of the principle is that any consideration of position of the electron in atomic and molecular species must be based in terms of probabilities, and is the reason for the development of the section of theoretical chemistry known as wave mechanics or quantum mechanics.

Problems

3.1 Is the model of a fixed, non-vibrating diatomic molecule consistent with the uncertainty principle? Your answer should be a justification for such molecules possessing zero-point vibrational energy.

3.2 Assuming an uncertainty of $\pm 5\%$ in the measurement of the velocity (and consequently momentum) of a car (mass = 10^6 g) travelling at about 30 m.sec^{-1} and of an electron travelling at about 10^6 m.sec^{-1}, calculate the uncertainty in their respective positions.

4

Quantum Mechanics

The purpose of this chapter is to introduce the reader to the subject of quantum mechanics or, as it is sometimes known, wave mechanics. The former description is used here since the latter one is somewhat ambiguous. The subject does not involve waves of any kind and is simply concerned with the calculation of the probabilities of location of the electrons in atoms and molecules, and the energies associated with such locations.

4.1 Energy Levels in the Hydrogen Atom

The equation representing the energy levels of the electron in the hydrogen atom as stated in Chapter 2 is repeated here:

$$E = -Rch/n^2 \qquad (4.1)$$

A pre-wave-mechanical derivation of equation (4.1) was carried out by Bohr in 1926, the details of which will not be given.

The Bohr model of the hydrogen atom was an electron circulating around the nucleus at a certain fixed radius r, as shown in Figure 4.1.

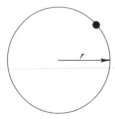

Figure 4.1. The Bohr model of the hydrogen atom

Such a model is not consistent with the uncertainty principle since Δq, the uncertainty in position, would be zero, and in consequence the uncertainty in momentum, Δp, would be infinite. The energy and, therefore, the momentum of the electron is known with reasonable accuracy. The smaller the uncertainty in the energy or momentum of the electron the larger is the uncertainty in its position. Models of the atom involving

Quantum Mechanics

fixed orbits in which electrons may circulate the nucleus are, therefore, not realistic.

Models of the atom involving electrons circulating about the nucleus are not even valid in terms of classical mechanics. Such models would involve the circulation of the electrons towards the nucleus in a decreasing spiral motion together with a continuous loss of energy. Neither of these processes have been observed in practice.

It should, however, be pointed out that in balancing the angular momentum of the electron with the attractive coulombic force in a model such as that in Figure 4.1 and by arbitrarily introducing quantization of energy, the correct energy level equation is obtained. Similar treatments of polyelectronic atoms, however, are very unsatisfactory.

4.2 The Schrödinger Equation

In the Bohr treatment of the atom quantization was introduced to the classical model, but the method of calculation which is now generally used is due to Schrödinger.

The basis of his method is to treat electrons as though their motions were governed by what is known as a *wave equation*. This does not mean that the electrons are to be considered as being waves or possessing any real wavemotion. It is an attempt to calculate probabilities of finding electrons with given values of their coordinates without implying anything concerning their physical nature. The Schrödinger method starts with a quantity, ψ, which is a function of the three spatial coordinates, x, y, and z, or sometimes more conveniently the polar coordinates, r, θ and ϕ, and also on the time. The difference between cartesian and polar coordinates is illustrated in Figure 4.2. The quantity, ψ, which we may call the wave function, can be written as:

$$\psi = f(x, y, z, t) \tag{4.2}$$

or

$$\psi = f(r, \theta, \phi, t) \tag{4.3}$$

Time-dependent wave functions describing electronic states are only of interest when an electron moves from one state to another, which occurs in electronic transitions. They are considered only as a basis for the understanding of electronic spectroscopy.

For the present treatment of atoms and molecules we will be interested mainly in the ground states which may be represented by time-independent wave functions, since the ground state (in the absence of any interactions) does not change with time.

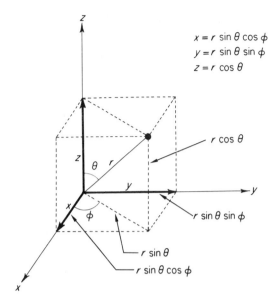

Figure 4.2. The relationship between cartesian and polar coordinates

To simplify the mathematics even further, we may consider a wave function which is unidimensional and depends only on the value of, say, the x coordinate. An equation which describes the standing waves (time-independent) of a stretched string is

$$\psi = A \sin \frac{2\pi x}{\lambda} \quad (4.4)$$

where A is the maximum amplitude of the wave function, x is the distance along the space coordinate in one dimension (direction), λ is the wavelength, and ψ is the amplitude of the wave for any given value of x.

By doubly differentiating ψ with respect to x we get

$$\frac{d\psi}{dx} = \frac{2\pi A}{\lambda} \cos \frac{2\pi x}{\lambda} \quad (4.5)$$

and

$$\frac{d^2\psi}{dx^2} = -\frac{4\pi^2 A}{\lambda^2} \sin \frac{2\pi x}{\lambda} \quad (4.6)$$

Quantum Mechanics

and eliminating $A \sin(2\pi x/\lambda)$ from equations (4.4) and (4.6) we get

$$\frac{d^2\psi}{dx^2} = -\frac{4\pi^2}{\lambda^2} \cdot \psi \qquad (4.7)$$

Now the kinetic energy of a particle is given by

$$T = \tfrac{1}{2}mv^2 \qquad (4.8)$$

which may be expressed in terms of the wavelength of the corresponding de Broglie 'wave' by writing the kinetic energy as

$$T = \tfrac{1}{2} \cdot \frac{(mv)^2}{m} = \frac{p^2}{2m} \qquad (4.9)$$

where $p = mv$ (momentum of the electron). Since

$$p = h/\lambda \text{ (see equation 3.7)} \qquad (4.10)$$

we may write

$$T = \frac{h^2}{2m\lambda^2} \qquad (4.11)$$

Eliminating λ^2 from equations (4.7) and (4.11) we obtain

$$\frac{d^2\psi}{dx^2} = -\frac{8\pi^2 mT}{h^2} \qquad (4.12)$$

The kinetic energy, T, may also be expressed in terms of the difference between the total energy of the system, E, and its potential energy, V:

$$T = E - V \qquad (4.13)$$

and equation (4.12) then becomes:

$$\frac{d^2\psi}{dx^2} = -\frac{8\pi^2 m}{h^2}(E - V) \cdot \psi \qquad (4.14)$$

which is usually rewritten in the form:

$$\frac{d^2\psi}{dx^2} + \frac{8\pi^2 m}{h^2}(E - V) \cdot \psi = 0 \qquad (4.15)$$

This is the Schrödinger equation for a particle of mass, m, total energy, E, potential energy, V, moving in one-dimensional space.

For the same particle moving in three-dimensional space, the equation becomes:

$$\frac{d^2\psi}{dx^2} + \frac{d^2\psi}{dy^2} + \frac{d^2\psi}{dz^2} + \frac{8\pi^2 m}{h^2}(E - V).\psi = 0 \qquad (4.16)$$

The quantity,

$$\frac{d^2}{dx^2} + \frac{d^2}{dy^2} + \frac{d^2}{dz^2}$$

may be abbreviated to ∇^2 where ∇ (del) is the operator

$$\frac{d}{dx} + \frac{d}{dy} + \frac{d}{dz}$$

An alternative method of writing the wave equation is

$$H\psi = E\psi \qquad (4.17)$$

where H is an operator known as the Hamiltonian operator. An operator is simply an instruction to carry out an operation upon whatever quantity it precedes. In the case of equation (4.17), H is operating on the quantity, ψ. For example, if H were 'add two' then $H\psi$ would mean 'add two to ψ' and the answer would be $2 + \psi$.

The appropriate operator in equation (4.17) is given by

$$H = -\frac{h^2}{8\pi^2 m} \cdot \nabla^2 + V \qquad (4.18)$$

which, when substituted into equation (4.17), gives equation (4.16)

The meaning of the operator form of the wave equation (e.g. equation 4.17) is that if ψ is operated upon by H, then the solution or solutions will contain expressions of ψ and energy values of the electronic system, E. These energy values are sometimes known as *eigenvalues* and the ψ functions as *eigenfunctions*.

If, for example, H was d/dx, the wave equation would become

$$\frac{d\psi}{dx} = E\psi \qquad (4.19)$$

A ψ function which satisfies equation (4.19) is

$$\psi = e^{nx} \qquad (4.20)$$

for

$$\frac{d\psi}{dx} = n\,e^{nx} \qquad (4.21)$$

Quantum Mechanics

and if n is called a quantum number, then the eigenvalues it represents are quantized as are the eigenfunctions, e^{nx}.

The eigenvalues are given by the value of n since if equations (4.20) and (4.21) are combined to read

$$\frac{d\psi}{dx} = n\psi \qquad (4.22)$$

and this equation is compared with equation (4.19), then we have

$$E = n \qquad (4.23)$$

which represents the energy values or eigenvalues associated with the various eigenfunctions, the latter being given by equation (4.20).

4.3 Interpretations of ψ

There are two ways of interpreting the meaning of ψ. The first of these is to consider the value of ψ as leading to a representation of electron density. This is to say that the electron is considered to be smeared out over the volume of an atom, and that $\psi^2 \, dx \cdot dy \cdot dz$ or $\psi^2 \, d\tau$ (where $d\tau = dx \cdot dy \cdot dz$) is a representation of the electron density in the volume element $d\tau$. (This is just the same as saying that the intensity of a beam of electromagnetic radiation is proportional to the square of the amplitude of the wave governing the motion of the photons—Section 1.6.) This was Schrödinger's original idea but Born suggested a more satisfactory interpretation which was that $\psi^2 \, d\tau$ was related to the probability of finding an electron in the volume element $d\tau$. The second interpretation is more satisfactory as it is easier to think of the electron in particulate terms rather than its being smeared out. The term electron density has, however, wide usage and it is better thought of as being a probability.

In probability theory, it is usual to assign unit probability to a *certain* event. Certainty is expressed as a probability of unity, being expressed as $P = 1$. If an event has zero probability, $P = 0$. The probabilities of other events may then be expressed as a number between zero and unity.

In quantum mechanics it is therefore convenient to arrange for the integral of the square of the eigenfunction with respect to a volume element, $d\tau$, to be unity. If the integral,

$$\int_0^\infty \psi^2 \, d\tau$$

is unity, then ψ is said to be normalized, an example of this being worked out in Section 4.6.

4.4 Limitations to the Nature and Form of ψ

There are three mathematical limitations upon the nature and form of ψ.

(1) The first of these is that ψ must be finite for all values of the coordinates. If, for instance, the value of ψ had an infinite value for a particular set of values of the space coordinates, it would result in there being an infinitely greater probability of finding the electron at that point than anywhere else. This violates the uncertainty principle. ψ must therefore have a finite or zero value at any point.

(2) The second limitation upon ψ is that it must be single-valued. This means that for a given set of values of the space coordinates there must be only one value of ψ leading to only one value of the probability of finding the electron with the particular values of the coordinates. To have more than one value of probability for the occurrence of an event is obviously ridiculous.

(3) The third limitation is that ψ should be continuous, i.e. should show no discontinuities but be a smoothly changing function.

The great significance of these limitations will be apparent in the next section. Without the limitations, quantization of energy levels would not appear as a feature of the solution of the wave equation.

It must also be pointed out that there is no proof available for the correctness or otherwise of the wave equation, or the limitation, mentioned above, upon the values of ψ, or indeed of the interpretation of ψ in terms of probability. What can be said is that there is a tremendous amount of evidence for their validity in the success of the treatment of many atomic and molecular problems.

4.5 Solutions of the Wave Equation for an Electron in a One-Dimensional Potential Well

One of the few cases in which it is possible to solve the wave equation exactly is for an electron which is confined to a one-dimensional potential well. A diagram of such a well is shown in Figure 4.3. The length along the x coordinate in which the electron can be found is l. For convenience we shall consider the electron to have zero potential energy, i.e. $V = 0$, inside the well.

The wave equation for the electron under these conditions is

$$\frac{d\psi^2}{dx^2} + \frac{8\pi^2 m}{h^2}[E]\psi = 0 \qquad (4.24)$$

Quantum Mechanics

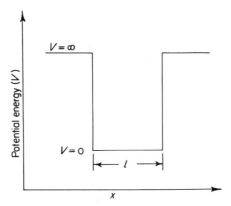

Figure 4.3. A one-dimensional potential well of length l

The eigenfunctions, ψ_n, must be such that ψ is zero for the region outside the well since we postulate that the electron is definitely in the well. They must also be such that the value of ψ is zero at the walls where $x = 0$ and $x = 1$. If this latter provision were not so, then there would be discontinuities at the boundaries. This point is demonstrated in Figures 4.4 and 4.5, where two situations are shown. Figure 4.4 shows a standing wave, where there are an integral number of half-wavelengths in the distance, l. In Figure 4.5 where the latter condition is not fulfilled there is no standing wave. The boundary conditions have already been referred to in Section 4.4.

Solutions of equation (4.24) are:

$$\psi_n = A \sin \frac{n\pi x}{l} \tag{4.25}$$

where ψ_n are the eigenfunctions, A is the maximum value of ψ, x is the distance along the x coordinate, l is the length of the potential well, and n is an integer (not zero). That equation (4.25) represents solutions of the equation (4.24) may be demonstrated by trial.

Equation (4.25) may be differentiated twice with respect to x, viz.:

$$\frac{d\psi}{dx} = \frac{n\pi}{l} \cdot A \cos \frac{n\pi x}{l} \tag{4.26}$$

and

$$\frac{d^2\psi}{dx^2} = -\frac{n^2\pi^2}{l^2} \cdot A \sin \frac{n\pi x}{l} \tag{4.27}$$

$$\frac{d^2\psi}{dx^2} = -\frac{n^2\pi^2}{l^2} \cdot \psi \qquad (4.28)$$

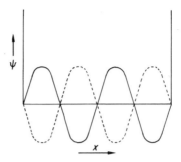

Figure 4.4. Standing wave

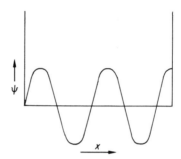

Figure 4.5. No standing wave

The results of this double differentiation may be used in conjunction with equation (4.24) to yield:

$$-\frac{n^2\pi^2}{l^2} \cdot A \sin\frac{n\pi x}{l} + \frac{8\pi^2 m}{h^2} \cdot E \cdot A \sin\frac{n\pi x}{l} = 0 \qquad (4.29)$$

or

$$-\frac{n^2\pi^2}{l^2} + \frac{8\pi^2 m}{h^2} \cdot E = 0 \qquad (4.30)$$

which may be rearranged to give an equation for the eigenvalues, E_n:

$$E_n = \frac{n^2 h^2}{8ml^2} \qquad (4.31)$$

As was the case with the eigenfunctions (equation 4.25) the integral number, n, appears in equation (4.31) for the eigenvalues. The essential

Quantum Mechanics

point to realize concerning n, is that only when there is an integral number of π units is ψ zero at $x = 0$ and $x = l$. It is the operation of the boundary condition which is responsible for the introduction of the number, n. Any non-integral number would not produce a standing wave. The integral, non-zero number n, is called a *quantum number* because it characterizes the quantization of the eigenvalues. It limits the electron to certain allowed energies.

The eigenfunctions, ψ_n, and the eigenvalues, E_n, are consistent with the originally set up wave equation, and the lower ones (for $n = 1, 2$ and 3) are shown in Figure 4.6, together with ψ_n^2.

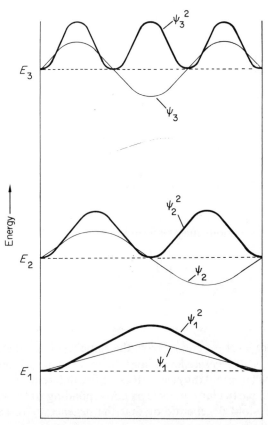

Figure 4.6. Plots of the eigenfunctions, ψ_n, and their squares, ψ_n^2, for $n = 1, 2$ and 3, together with an indication of their respective eigenvalues (energies)

The important conclusions from this wave mechanical treatment of the electron in a one-dimensional potential well are summarized below.

(1) Quantization does not have to be built into the mathematics, but arises as a result of the boundary conditions.
(2) As far as the specific problem under consideration is concerned, there are some important generalizations about the eigenvalues, which are that E increases as the square of the quantum number and decreases as the inverse square of the length of the well.

These latter points are of great importance when considering the electronic configurations of certain classes of molecules, and further discussion will be found in Sections 8.4 and 9.7.

4.6 Normalization of the Eigenfunctions

For ψ_n to be normalized we have to ensure that

$$\int_0^\infty \psi_n^2 \, dx = 1\cdot 0 \tag{4.32}$$

Putting our value for ψ_n in this integral and equating its value to unity we obtain:

$$\int_0^\infty A^2 \sin^2 \frac{n\pi x}{l} \cdot dx = 1\cdot 0 \tag{4.33}$$

The value of A which satisfies equation (4.33) is $(2/l)^{1/2}$ so that the normalized wave functions may be written as:

$$\psi_n = \left(\frac{2}{l}\right)^{1/2} \sin \frac{n\pi x}{l} \tag{4.34}$$

4.7 Degeneracy

If two or more levels of energy have the same value for their energies they are said to be *degenerate* levels, and the number of such levels is the *degeneracy*. For example, if the eigenvalues for five different eigenfunctions are identical, the particular energy level corresponding to the eigenvalue is said to be five-fold degenerate or that the degeneracy of the level is five-fold.

The occurrence of the degeneracy of energy levels may be illustrated by considering the solution of the wave equation for an electron in a two-

Quantum Mechanics

dimensional potential well. Such a well is shown in Figure 4.7. The appropriate wave equation is

$$\frac{d^2\psi}{dx^2}+\frac{d^2\psi}{dy^2}+\frac{8\pi^2 m}{h^2}\cdot E\cdot\psi = 0 \qquad (4.35)$$

since $V = 0$ by definition.

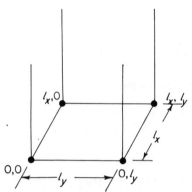

Figure 4.7. A two-dimensional potential well

The method of solution of equation (4.35) involves the separation into two parts, the one part being concerned with variation of ψ along the x coordinate and the other with its variation along the y coordinate.

We can write the total wave function, $\psi_{x,y}$, as the product of two wave functions, ψ_x and ψ_y, these latter being dependent only upon x and y respectively:

$$\psi_{x,y} = \psi_x \cdot \psi_y \qquad (4.36)$$

If this is so, then because ψ_y is independent of x, we can write:

$$\frac{d^2\psi_{x,y}}{dx^2} = \psi_y \cdot \frac{d^2\psi_x}{dx^2} \qquad (4.37)$$

and since ψ_x is independent of y, we can write:

$$\frac{d^2\psi_{x,y}}{dy^2} = \psi_x \cdot \frac{d^2\psi_y}{dy^2} \qquad (4.38)$$

Substituting in the wave equation, (4.35), we have

$$\psi_y \cdot \frac{d^2\psi_x}{dx^2}+\psi_x \cdot \frac{d^2\psi_y}{dy^2}+\frac{8\pi^2 m}{h^2}\cdot E \cdot \psi_x \cdot \psi_y = 0 \qquad (4.39)$$

We can divide equation (4.39) throughout by the factor, $(8\pi^2 m)/h^2 \cdot \psi_x \cdot \psi_y$, and arrive at

$$\frac{h^2}{8\pi^2 m} \cdot \frac{1}{\psi_x} \cdot \frac{d^2 \psi_x}{dx^2} + \frac{h^2}{8\pi^2 m} \cdot \frac{1}{\psi_y} \cdot \frac{d^2 \psi_y}{dy^2} = -E \qquad (4.40)$$

We can achieve separation of the variables of equation (4.40) by considering E to be made up from separate contributions E_x and E_y such that

$$E = E_x + E_y \qquad (4.41)$$

in which case equation (4.40) becomes

$$\frac{h^2}{8\pi^2 m} \cdot \frac{1}{\psi_x} \cdot \frac{d^2 \psi_x}{dx^2} + \frac{h^2}{8\pi^2 m} \cdot \frac{1}{\psi_y} \cdot \frac{d^2 \psi_y}{dy^2} = -E_x - E_y \qquad (4.42)$$

which is the sum of the equations (4.43) and (4.44):

$$\frac{h^2}{8\pi^2 m} \cdot \frac{1}{\psi_x} \cdot \frac{d^2 \psi_x}{dx^2} = -E_x \qquad (4.43)$$

$$\frac{h^2}{8\pi^2 m} \cdot \frac{1}{\psi_y} \cdot \frac{d^2 \psi_y}{dy^2} = -E_y \qquad (4.44)$$

The normalized solutions of these latter two equations are:

$$\psi_x = \left(\frac{2}{l_x}\right)^{1/2} \sin \frac{n_x \pi}{l_x} \qquad (4.45)$$

and

$$\psi_y = \left(\frac{2}{l_y}\right)^{1/2} \sin \frac{n_y \pi}{l_y} \qquad (4.46)$$

which involves two quantum numbers n_x and n_y due to the imposition of the relevant boundary conditions.

The eigenvalues are also given by two equations, these being

$$E_x = \frac{n_x^2 h^2}{8m l_x^2} \qquad (4.47)$$

and

$$E_y = \frac{n_y^2 h^2}{8m l_y^2} \qquad (4.48)$$

Quantum Mechanics

The total energy of the system is given by equation (4.41) which may now be written in the form:

$$E = \frac{h^2}{8m}\left(\frac{n_x^2}{l_x^2} + \frac{n_y^2}{l_y^2}\right) \qquad (4.49)$$

The allowed energy levels depend now on the combination of the values of the two quantum numbers, n_x and n_y, as can be demonstrated for the first few values shown in Table 4.1.

Table 4.1. Dependence of the Total Energy of an Electron in a Two-dimensional Potential Well on the Values of the Quantum Numbers, n_x and n_y

n_x	n_y	Total energy (units of $h^2/8m$)
1	1	$\frac{1}{l_x^2} + \frac{1}{l_y^2}$
2	1	$\frac{4}{l_x^2} + \frac{1}{l_y^2}$
1	2	$\frac{1}{l_x^2} + \frac{4}{l_y^2}$
2	2	$\frac{4}{l_x^2} + \frac{4}{l_y^2}$

For the four combinations of the values, 1 and 2, for the two quantum numbers we obtain four different values for the total energy of the system. These energies depend on the dimensions of the potential well and degeneracy arises when $l_x = l_y$, in which case the four combinations above give rise to only three different values for the total energy since

$$\frac{4}{l_x^2} + \frac{1}{l_y^2} = \frac{1}{l_x^2} + \frac{4}{l_y^2} \qquad (4.50)$$

This particular level is now doubly degenerate and in general the more symmetrical the system the more widespread is the phenomenon of degeneracy. This generalization has important consequences in the study of polyatomic molecules and polyelectronic atoms.

4.8 Dimensions and Quantum Numbers

A calculation similar to that carried out in Section 4.7 may be done for the case of a three-dimensional potential well, and it is not difficult to see that the solutions of the wave equation will depend upon the values of *three* quantum numbers. This is an important and general conclusion since it means that the energy levels available to the electrons in three-dimensional atoms can be described adequately by the use of three quantum numbers.

4.9 Non-Zero Values of the Quantum Numbers

It has been pointed out that the quantum numbers used for the description of electronic energy levels may possess integral but non-zero values. If n in equation (4.25) had the value zero, then the factor, $n\pi x/l$, would at all values of x be zero, and ψ would also be zero indicating zero probability of finding the electron in the potential well. This, we know, is not so because we have postulated its presence in the well and $\psi = 0$ at any point outside the well. The consequences of this are that the electron must exist in the well but that there is zero probability of it existing in the level, $E_{n=0}$.

Considering the situation in terms of the Heisenberg Uncertainty Principle, we could say that since the electron was in the well, the uncertainty in its position is given by the length of the well, l_x, i.e.

$$\Delta q = l_x \tag{4.51}$$

The uncertainty in its momentum is therefore

$$\Delta p \sim \frac{h}{l_x} \tag{4.52}$$

and the uncertainty in the kinetic energy is given by

$$\Delta T \sim \frac{(\Delta p)^2}{2m} \sim \frac{h^2}{2ml_x^2} \tag{4.53}$$

(see equation 4.9), so that the total energy ($V = 0$) must be of the same order as that given by equation (4.53). The lowest energy which is consistent with this value is that obtained by putting $n = 1$ in equation (4.31) which gives

$$E = \frac{h^2}{8ml_x^2} \tag{4.54}$$

Quantum Mechanics

4.10 Conclusions

The wave equation has been derived for up to three-dimensional systems and has been solved by inspection for electrons in several potential wells. Normalization of eigenfunctions and degeneracy of eigenvalues have been demonstrated. We are now in a position to apply quantum mechanics to some atomic systems.

Problems

4.1 Derive an expression for the eigenvalues of an electron in a three-dimensional potential well which has sides with the lengths, l_x, l_y and l_z.

4.2 Using the expression derived in the previous problem calculate the energies of the levels which involve quantum number values of 1 and 2. If

(a) $l_x = l_y \neq l_z$ and
(b) $l_x = l_y = l_z$,

what effects are there upon the degeneracy of the various levels?

5

The Application of Quantum Mechanics to Atomic Problems

5.1 The Eigenfunctions and Eigenvalues for Hydrogen-like Atoms

A hydrogen-like atom is one in which the nucleus of charge $+Ze$ is accompanied by a single extranuclear electron. Examples of hydrogen-like atoms are H, He^+, Li^{2+}, Be^{3+}, B^{4+} and C^{5+} which all have the electronic configuration $1s^1$ in their ground states. The wave equation for such species is capable of exact solution. The wave equation may now be written as:

$$\nabla^2 \psi + \frac{8\pi^2 \mu}{h^2}\left(E + \frac{Ze^2}{r}\right)\psi = 0 \qquad (5.1)$$

since the potential energy of the system is given by

$$V = -\frac{Ze^2}{r} \qquad (5.2)$$

the zero of energy being that for the infinite separation of the nucleus of charge, $+Ze$, and the electron of charge, $-e$.

In equation (5.1) the symbol μ replaces m for the mass of the electron where μ is the reduced mass of the electron, i.e.

$$\mu = \frac{mM}{m+M}$$

M being the mass of the nucleus. Since $M \gg m$ then $m+M \sim M$ and therefore $\mu \sim m$.

The equation as written is in terms of cartesian coordinates but the solutions are best obtained in terms of polar coordinates (Section 4.2).

The total wave function, $\psi_{r,\theta,\phi}$, may be separated into three parts, ψ_r, ψ_θ and ψ_ϕ, these being solved separately, and the solutions being recombined to give the solutions of the total wave equation.

The Application of Quantum Mechanics to Atomic Problems

The results of such a treatment may be summarized as follows:

(1) Three quantum numbers are essential for the description of the total wave function. These are designated by the letters, n, l and m (Section 4.8).
(2) The complete eigenfunctions are expressed as the product of a radial factor, ψ_r, and the angular factors, ψ_θ and ψ_ϕ, i.e.

$$\psi_{r,\theta,\phi} = \psi_r \cdot \psi_\theta \cdot \psi_\phi \qquad (5.3)$$

(3) The radial factor depends on the values of the two quantum numbers, n and l. The *principal quantum number* is n and has the permitted values of 1, 2, 3, 4,..., etc., whereas the value of l, the *orbital angular momentum quantum number*, depends on the value of n in any given situation. The quantum number, l, may be 0, 1, 2, 3 up to a *maximum value of* $n-1$ for any value of n.
(4) The angular factors, ψ_θ and ψ_ϕ, depend upon the values of l, m and m respectively. The *magnetic quantum number*, m, has the values, l, $l-1$, $l-2$, $\cdots$ 0, $\cdots$ $-(l-2)$, $-(l-1)$, $-l$, for any value of l. There are $2l+1$ values of m for any value of l.
(5) The eigenvalues are given by

$$E = -\frac{2\pi^2 \mu Z^2 e^4}{n^2 h^2} \qquad (5.4)$$

The values of energy for any finite value of n are negative. The reason for this is that the energy values are expressed relative to a zero of energy which is taken to be the energy of the completely separated electron and nucleus. Equation (5.4) expresses the energy change when the electron and the nucleus are brought together from infinity to one of the many quantized states. These energy changes are negative in accordance with the convention concerning exothermic changes (i.e. changes where energy is released).

For a given value of n there are a number of degenerate eigenfunctions, the eigenvalue being independent of the values of l and m. For $n = 1$ the only permitted value of l is zero and consequently m is zero so that the degeneracy of the eigenfunction is unity. For $n = 2$ we have $l = 1$, $m = +1, 0, -1$, giving three different levels, and $l = 0$, $m = 0$, making four degenerate levels in all. For $n = 3$ there are nine levels and, in general, the degeneracy of any one eigenvalue is given by n^2 for hydrogen-like atoms (this is not the case for any other atoms). A diagram of some of these levels is given in Figure 5.1 and the corresponding eigenfunctions are shown in Table 5.1.

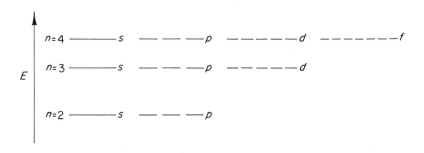

Figure 5.1. Some of the eigenvalues for a hydrogen-like atom, showing their degeneracies

Table 5.1. The Forms of the Radial and Angular Wave Functions for Hydrogen-like Atoms

Orbital	Radial function (ψ_r)	Angular function ($\psi_{\theta,\phi}$)
$1s$	$AZ^{3/2}\,e^{-kZr/2}$ [a]	B
$2s$	$CZ^{3/2}\,(2-e^{kZr/2})\,e^{-kZr/4}$	B
$2p_x$	$DZ^{5/2}\,kr\,e^{-kZr/4}$	$E \sin\theta \cos\phi$
$2p_y$	$DZ^{5/2}\,kr\,e^{-kZr/4}$	$E \sin\theta \sin\phi$
$2p_z$	$DZ^{5/2}\,kr\,e^{-kZr/4}$	$E \cos\theta$
$3d_{z^2}$	$FZ^{7/2}\,k^2r^2\,e^{-kZr/9}$	$G(3\cos^2\theta - 1)$
$3d_{x^2-y^2}$	$FZ^{7/2}\,k^2r^2\,e^{-kZr/9}$	$H \sin^2\theta \cos 2\phi$
$3d_{xy}$	$FZ^{7/2}\,k^2r^2\,e^{-kZr/9}$	$H \sin^2\theta \sin 2\phi$
$3d_{xz}$	$FZ^{7/2}\,k^2r^2\,e^{-kZr/9}$	$H \sin\theta \cos\theta \cos\phi$
$3d_{yz}$	$FZ^{7/2}\,k^2r^2\,e^{-kZr/9}$	$H \sin\theta \cos\theta \sin\phi$

[a] The symbols k and A to H signify constant terms.

5.2 Atomic Orbitals

The solution of the wave equation for hydrogen-like atoms gives the allowed eigenfunctions which enable a calculation of the probabilities of finding electrons with particular values of the three space coordinates to be made. The eigenfunctions given in Table 5.1 are normalized (Section 4.6) and therefore the square of the wave function gives the probability we require. It is common to calculate the volume in which there is a 95 per cent chance or 0·95 probability of finding the electron. Such

The Application of Quantum Mechanics to Atomic Problems

volumes within the hydrogen-like atom are known as *atomic orbitals*. Strictly speaking the *eigenfunctions* themselves are the *atomic orbitals* but it is useful to think of them as being represented by the volumes previously mentioned. Several generalizations are helpful in talking about atomic orbitals. The first is their coding in terms of the value of l. This is shown in Table 5.2.

Table 5.2. Coding of Atomic Orbitals

Value of l	Code letter
0	s
1	p
2	d
3	f
4	g
5	h

The limitations upon the value of l, as given in Section 5.1, make possible the following atomic orbitals in a hydrogen-like atom:

1s
2s 2p
3s 3p 3d
4s 4p 4d 4f
5s 5p 5d 5f 5g
6s 6p 6d 6f 6g 6h, etc.

It will be noticed that these descriptions of atomic orbitals do not involve m, the magnetic quantum number. This is because in the absence of a magnetic field all orbitals of a given n value are degenerate, but the value of m tells us the degeneracy $(2l+1)$ of any set of orbitals with the same l value. The value of m determines the number of spatially different atomic orbitals with identical values of n and l. For an l value of zero there can only be one orbital (s), for $l = 1$ there may be three (p), corresponding to m values of $+1$, 0 and -1, while for $l = 2$ there are five possible values of m, viz.: $+2$, $+1$, 0, -1 and -2. This situation is generalized in Table 5.3, so that there is

one s orbital for each value of n
three p orbitals for each value of n which is $\geqslant 2$
five d orbitals for each value of n which is $\geqslant 3$
seven f orbitals for each value of n which is $\geqslant 4$

etc.

Table 5.3. Dependence of Number of Atomic Orbitals upon the Value of n

Value of n	Number of atomic orbitals					
	s	p	d	f	g	
1	1					
2	1	3				
3	1	3	5			
4	1	3	5	7		
5	1	3	5	7	9	etc.

It is useful to have a pictorial representation of concepts such as atomic orbitals which are in reality only mathematical equations. We can do this for the radial and angular parts of the eigenfunction for the hydrogen-like atomic orbitals.

Consider the radial function of the 1s orbital. A plot of ψ_r as a function of the distance from the nucleus is shown in Figure 5.2.

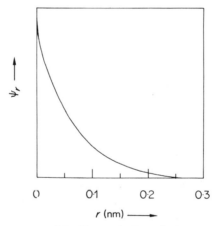

Figure 5.2. The variation of the radial function of the 1s atomic orbital of the hydrogen atom with the distance from the nucleus

The function is spherically symmetrical so that the situation represented in Figure 5.1 could be along any of the infinite number of radii of the hydrogen-like atom. Since the square of ψ is related to the probability density of finding the electron at any given point, Figure 5.2 would indicate that the maximum probability would occur at the nucleus

The Application of Quantum Mechanics to Atomic Problems

($r = 0$). That this is not so is demonstrated by calculating the *radial probability distribution function*. This is done by calculating the probability of finding the electron at a given distance from the nucleus, taking all directions into account. We can calculate the probability of finding the electron in a volume element which varies in radius between r and $r+dr$ as follows:

$$\tfrac{4}{3}\pi(r+dr)^3 - \tfrac{4}{3}\pi r^3 = \tfrac{4}{3}\pi r^3 + 4\pi r^2\, dr + 4\pi r(dr)^2$$
$$+ \tfrac{4}{3}\pi(dr)^3 - \tfrac{4}{3}\pi r^3$$
$$= 4\pi r^2\, dr \text{ (if second and third order } dr \text{ terms are neglected)}$$

The radial probability distribution function is

$$R = \psi_r^2 \times 4\pi r^2 \qquad (5.5)$$

and a plot of this against r for the 1s atomic orbital is shown in Figure 5.3.

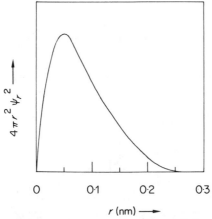

Figure 5.3. The variation of the radial probability distribution function of the 1s atomic orbital of the hydrogen atom with the distance from the nucleus

As can be seen from this figure, the distance from the nucleus at which there is maximum probability of finding the electron is 0·053 nm. The area under these plots is unity for normalized functions, and although R_{1s} is only zero at $r = \infty$ it can be seen that the majority of the probability of finding the electron is within a much shorter distance from the nucleus.

The radial eigenfunctions and the corresponding radial probability distribution functions are shown in Figure 5.4 for the 2s, 2p, 3d and 4f orbitals.

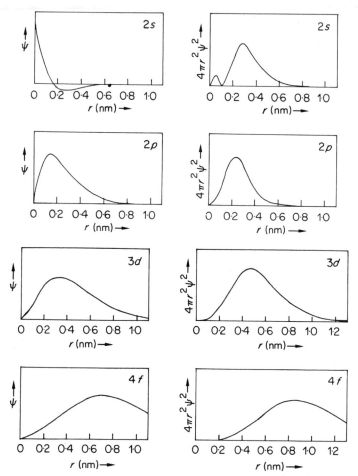

Figure 5.4. Plots of the radial eigenfunctions and radial probability distribution functions of the 2s, 2p, 3d and 4f orbitals of the hydrogen atom

It is possible to interpret the radial and angular wave functions in a pictorial manner by constructing what are known as boundary surfaces for each type of atomic orbital. A boundary surface encloses a volume in which there is a 0·95 probability of finding the electron. The figure of 0·95 is an arbitrary choice and in the case of the hydrogen atom (for the 1s orbital) leads to a boundary surface which is a sphere (with a radius of about 0·2 nm) as is shown in Figure 5.5.

Orbitals with other than s character are not spherically symmetrical

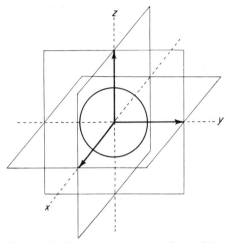

Figure 5.5. The 1s boundary surface of the hydrogen atom

and some further explanation is required. Consider the three degenerate p eigenfunctions which may be written as:

$$\psi_{l=1, m=+1} = A \cdot \sin \theta \, e^{+i\phi} \quad (5.6)$$

$$\psi_{l=1, m=0} = A \cdot \cos \theta \quad (5.7)$$

$$\psi_{l=1, m=-1} = A \cdot \sin \theta \, e^{-i\phi} \quad (5.8)$$

For m values of $+1$ or -1 the eigenfunction contains $i, (\sqrt{-1})$ and therefore are imaginary functions which cannot be drawn or visualized in physical terms. The $\psi_{1,0}$ eigenfunction (equation 5.7) is symmetrical about the z axis and is known as the p_z orbital.

By a mathematical transformation we can produce two other orbitals which are real and can lead to a pictorial method of expression. It is known that if ψ_A and ψ_B are eigenfunctions of the equation, $H\psi = E\psi$, then any linear combination of ψ_A and ψ_B will be an eigenfunction. The linear combinations, $\psi_A + c\psi_B$ and $\psi_A - c\psi_B$, will be eigenfunctions of the original equation and if we write $\psi_A = \psi_{1,1}$ and $\psi_B = \psi_{1,-1}$ and $c = 1$, then we have two new eigenfunctions which are symmetrical with respect to the x and y axes respectively (see Figure 4.2):

$$\psi_x = \psi_{1,1} + \psi_{1,-1}$$
$$= A \sin \theta \, e^{+i\phi} + A \sin \theta \, e^{-i\phi}$$
$$= A \sin \theta [e^{i\phi} + e^{-i\phi}]*$$
$$= A \sin \theta \cos \phi \quad (5.9)$$

and

$$\psi_y = \psi_{1,1} - \psi_{1,-1}$$
$$= A \sin \theta \, e^{+i\phi} - A \sin \theta \, e^{-i\phi}$$
$$= A \sin \theta [e^{i\phi} - e^{-i\phi}]*$$
$$= [e^{i\phi} - e^{-i\phi}]$$
$$= 2i \, A \sin \theta \sin \phi \tag{5.10}$$

Equations (5.9) and (5.10) when normalized may be written as

$$\psi_x = N \sin \theta \cos \phi \tag{5.11}$$

and

$$\psi_y = N \sin \theta \sin \phi \tag{5.12}$$

respectively, a factor of i^{-1} being introduced into equation (5.10) in the normalization procedure to eliminate i. The functions which have been represented as ψ_x and ψ_y (equations 5.11 and 5.12) are symmetrical with respect to the x and y axes respectively, and are known as the p_x and p_y atomic orbitals.

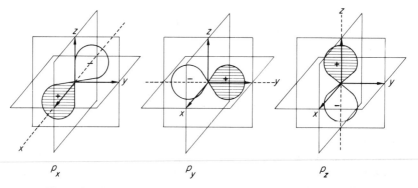

Figure 5.6. Angular dependence of the p_x, p_y and p_z atomic orbitals

* The quantities in parentheses are converted into trigonometric functions by using Euler's identities:

$$e^{i\phi} = \cos \phi + i \sin \phi$$

and

$$e^{-i\phi} = \cos \phi - i \sin \phi$$

The Application of Quantum Mechanics to Atomic Problems

The distributions of the angular parts of the three p orbitals are shown in Figure 5.6. The $+$ and $-$ signs, which are very important when bonding between atoms is considered, indicate the sign of the eigenfunction in the various regions of the orbitals. If the points (x, y, z) and $(-x, -y, -z)$ are compared for an s orbital it is found that the sign and magnitude of ψ are the same. A similar comparison for a p orbital shows that while the magnitude of ψ is the same at the two points, its sign differs: at one point ψ is positive and at the other ψ is negative.

The boundary surfaces for the p orbitals are not shown but have the same general shape and the same spatial orientation as the angular dependences of those orbitals. The boundary surfaces are based upon ψ^2 values where ψ^2 is positive in all regions of space. As an approximation the shapes of the angular dependences of orbitals may be taken as a representation of the shapes of the orbitals.

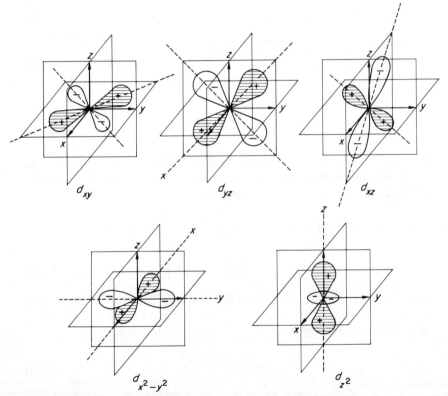

Figure 5.7. Angular dependence of d_{xy}, d_{yz}, d_{xz}, $d_{x^2-y^2}$ and d_{z^2} orbitals

For a value of $l = 2$ there are five degenerate (d) orbitals which differ in their spatial arrangement, as is shown in Figure 5.7. They are distinguished by the cartesian subscripts which indicate the axes which are important in their descriptions. Unlike the p orbitals the shapes of the d orbitals differ, there being three different types. In fact they are all equivalent and are so arranged that they can be drawn. By a similar procedure as that used for the p orbitals the imaginary functions have been combined in various linear combinations to give real functions. This means that the five d orbitals must not be associated with specific values of the magnetic quantum number.

In future discussions the eigenfunctions referred to will be described as s or p_x or d_{z^2}, etc. It must be emphasized that the total eigenfunction is the product of the radial function and the angular function.

5.3 The Eigenvalues of the Hydrogen-like Atomic Orbitals

Each atomic orbital has a corresponding eigenvalue given by equation (5.4) and depends only on the value of the principal quantum number, n. It is interesting to note that equation (5.4) explains in detail the emission spectrum of hydrogen and hydrogen-like atoms. The most stable (i.e. lowest energy) level or eigenvalue is that for which $n = 1$ and which has the energy,

$$E_1 = -\frac{2\pi^2 \mu Z^2 e^4}{h^2}$$

All other values of n correspond to levels of higher energy, and if an electronic transition occurs from such a level to the lowest one, the energy emitted would be expected to be given by

$$E_n - E_1 = -\frac{2\pi^2 \mu Z^2 e^4}{n^2 h^2} + \frac{2\pi^2 \mu Z^2 e^4}{h^2}$$

$$= \frac{2\pi^2 \mu Z^2 e^4}{h^2}\left(1 - \frac{1}{n^2}\right) \tag{5.13}$$

This has precisely the same form as the equation (2.12) derived empirically for the Lyman lines of the emission spectrum of hydrogen. Equations (2.12) and (5.13) are identical if

$$Rch = \frac{2\pi^2 \mu Z^2 e^4}{h^2} \tag{5.14}$$

and since, for hydrogen, $Z = 1$, a value of the Rydberg constant, R, in

The Application of Quantum Mechanics to Atomic Problems

units of cm^{-1}, may be obtained by evaluating the expression for

$$R = \frac{2\pi^2 \mu e^4}{ch^3} \quad (5.15)$$

If the known values of μ, e, c, h and π are substituted into equation (5.15), a value for R is obtained which is in extremely close agreement with the experimentally determined value. It is important to notice that the value of the so-called Rydberg *constant* varies with μ but is constant for any one hydrogen-like atom. The variation in μ over the known elements amounts to only 0·06 per cent of the total value of R, and for general purposes may be disregarded. For hydrogen-like atoms in general, ignoring the variation of R with μ, we may write for the energy levels:

$$E = -\frac{RchZ^2}{n^2} \quad (5.16)$$

so that for a transition from n_2 to n_1 the energy released will be

$$E_2 - E_1 = -\frac{RchZ^2}{n_2^2} + \frac{RchZ^2}{n_1^2}$$

$$= RchZ^2 \left(\frac{1}{n_1^2} - \frac{1}{n_2^2} \right) \quad (5.17)$$

which can be used as a general expression for the emission (or absorption) spectra of such atoms.

The ionization potentials of a series of hydrogen-like atoms are known. The transition corresponding to ionization from the lowest electronic level will be that where $n_1 = 1$ and $n_2 = \infty$, so that

$$I = E_\infty - E_1 = RchZ^2 \left(\frac{1}{1^2} - \frac{1}{\infty^2} \right)$$

$$= RchZ^2 \quad (5.18)$$

Equation (5.18) accurately reproduces the values of the ionization potentials of hydrogen-like atoms which can be obtained by experimental means.

5.4 Polyelectronic Atoms

Hydrogen-like atoms have only one electron surrounding the nucleus and the wave equation is capable of exact solution for such systems. This is not the case for polyelectronic atoms which have more than one extranuclear electron. It is not possible to solve the wave equation exactly for the helium atom in which there are three particles to consider. The wave equation may be written as:

$$\nabla_1^2\psi + \nabla_2^2\psi + \frac{8\pi^2\mu}{h^2}\left(E + \frac{2e^2}{r_1} + \frac{2e^2}{r_2} - \frac{e^2}{r_{12}}\right)\psi = 0 \quad (5.19)$$

r_1 and r_2 being the distances of the two electrons respectively from the nucleus and r_{12} being the interelectronic distance. There are as many del squared terms as there are electrons. Since an exact solution of the simplest polyelectronic wave equation is not possible, several approximate methods have been developed. It is not proposed to deal with them in this book, except to mention two of them.

The underlying idea that a system will normally exist in its lowest energy state is the basis of the *variation principle*. For a given form of an approximate wave function the variation principle allows one to optimize it such that the best function of the particular form has the lowest associated energy. For example, if an approximate wave function is

$$\psi = e^{-cr} \quad (5.20)$$

expressing the variation of ψ with distance from the nucleus, r, the variation principle allows one to obtain a 'best value' for the term, c. In practice the energy (eigenvalues) associated with the function (5.20) is calculated for various values of c. A plot of this associated energy against the value of c may be as shown in Figure 5.8.

The value of c obtained by this method gives the best eigenfunction of the form of equation (5.20). It does not eliminate the possibility of there being some eigenfunction of *different* form which when optimized in a similar manner will give an even lower associated energy. However, after many trials and errors, something approaching the true eigenfunction may be obtained.

The approximate solution of the wave equation for a polyelectronic system can be carried out by the *Self-Consistent Field* (SCF) method. The basis of the method is that an approximate set of atomic orbitals is chosen for a given system. From these an average potential acting upon each electron is calculated. These potentials are then used to calculate new orbitals from which a better approximation to the average potential can be calculated. The process is repeated until a set of orbitals reproduces the potential which gave rise to those orbitals. The orbitals (wave functions) resulting from this type of treatment are similar to the hydrogen orbitals and may be given the same labels. They are known as SCF orbitals.

The Application of Quantum Mechanics to Atomic Problems

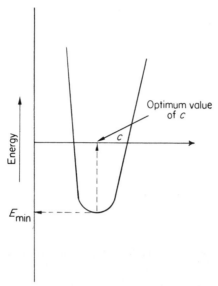

Figure 5.8. A plot of energy against c for the function, $\psi = e^{-cr}$, showing the operation of the variation principle in obtaining the best value of c

5.5 Orbital Penetration Effects

For hydrogen-like atoms all orbitals with the same value of the principal quantum number are degenerate. This fact simplifies the emission spectroscopy of these species considerably. For polyelectronic atoms there occurs a breakdown of this degeneracy due to what are known as orbital penetration effects. A suitable example would be to consider the atom of lithium. The 1s orbital is occupied by two electrons and the radial distribution function is shown in Figure 5.9, together with the superimposed functions for the 2s and 2p orbitals.

In spite of the main maximum of the 2p function being nearer the nucleus than that of the 2s function, it is the latter which has the greater probability of interacting with the electron core of the lithium atom. The nearer the nucleus the electron is the less is the shielding it receives from other electrons and the lower is its energy. This produces the breakdown in degeneracy between the 2s and 2p orbitals, the former being a more stable orbital in a polyelectronic atom than the latter.

Extension of these arguments applies to the breaking of the degeneracy of the $n = 3$ levels. In Figure 5.10 is shown the radial probability

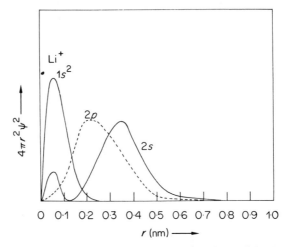

Figure 5.9. The radial distribution functions of the 2s and 2p orbitals of the lithium atom together with that of the lithium core ($1s^2$)

distribution functions for the ten-electron sodium core and those for the 3s, 3p and 3d orbitals.

It can be seen that the 3s orbital 'penetrates' further into the core than the 3p orbital, and that the 3d orbital hardly penetrates at all. Since the core of ten electrons will shield any electrons outside it from the effect of the nucleus, it means that the 3d orbitals will have maximum shielding

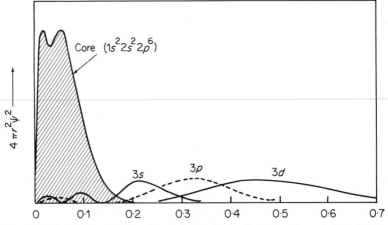

Figure 5.10. The radial distribution functions of the 3s, 3p and 3d orbitals together with that of the sodium core ($1s^2\ 2s^2\ 2p^6$)

The Application of Quantum Mechanics to Atomic Problems

followed by the 3p orbitals, and the 3s orbital will receive least shielding. Consequently an electron in a 3s orbital will be more stable than an electron in a 3p orbital.

The energies or stabilities of SCF orbitals of polyelectronic atoms vary with the nuclear charge (cf. equation 5.4). The energy of the $n = 1$ level in helium is four times less than the same level in hydrogen according to equation (5.4). However, the ionization potential of helium is 24 ev which is just less than twice that of hydrogen (13·6 ev). This is due to electron shielding or, to put it another way, due to interelectronic repulsion effects. If the explanation is due to one electron shielding the other from the nucleus, we may use equation (5.18) to write:

$$I_H = RchZ^2 = Rch \qquad (5.21)$$

so that

$$I_{He} = I_H Z^2 \qquad (5.22)$$

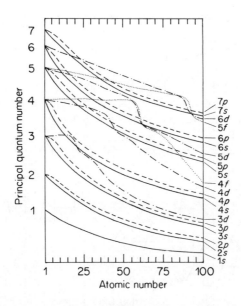

Figure 5.11. A diagrammatic indication of the variation of orbital energies with increasing nuclear charge (qualitative). (From page 629 of *Advanced Inorganic Chemistry*, by F. A. Cotton and G. Wilkinson, 2nd Edition, Interscience; reproduced by permission of the publishers)

If, in equation (5.22), Z is replaced by Z_{eff} representing the effective nuclear charge of the helium nucleus as far as the electron to be ionized is concerned, we get

$$I_{He} = I_H Z_{eff}^2 \qquad (5.23)$$

Putting in the observed values for I_H and I_{He} we have

$$Z_{eff} = \left(\frac{I_{He}}{I_H}\right)^{1/2} = 1\cdot33 \qquad (5.24)$$

Thus one electron contributes a shielding effect of $2 - 1\cdot33 = 0\cdot67$ electronic charge units to the attraction of the nucleus for the other electron.

A diagram of the variation of orbital energies with increasing nuclear charge is shown in Figure 5.11.

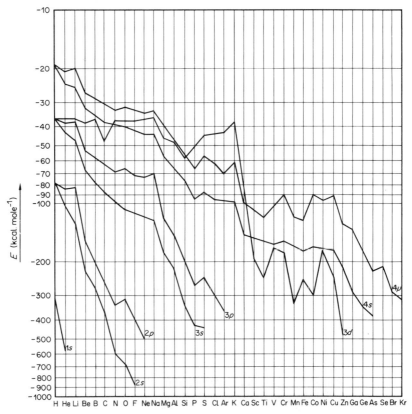

Figure 5.12. The variation of orbital energy with atomic number for the $1s$, $2s$, $2p$, $3s$, $3p$, $3d$, $4s$ and $4p$ orbitals

The Application of Quantum Mechanics to Atomic Problems

It will be noticed that the order of energies of the orbitals varies with Z and that levels of different n values overlap. This is of great significance in the understanding of the electronic configurations of the elements and consequently in the understanding of the chemistry of the elements.

The diagram in Figure 5.11 is only qualitative as far as the relative energies of the various orbitals are concerned. A quantitative plot of orbital energies versus atomic number is shown in Figure 5.12 for the 1s, 2s, 2p, 3s, 3p, 3d, 4s and 4p orbitals as far as measurements allow. The effects of orbital penetration on the energy of the 4s level are obvious, and also apparent is the general stabilizing effect of increasing the nuclear charge.

5.6 Summary

The wave equation may be solved exactly for hydrogen-like atoms. The solutions involve eigenvalues or energy levels depending on the value of n, the principal quantum number, and eigenfunctions corresponding to these eigenvalues (the latter being degenerate to the extent of n^2). The eigenfunctions allow calculation of the regions within the atoms where electrons may be found. These regions described by n and l are known as atomic orbitals. The number of different atomic orbitals with the same n and l values is given by $2l+1$, which is also the number of different values of the magnetic quantum number, m.

For polyelectronic atoms we can still talk about atomic orbitals and give them the same labels. Those orbitals with the same n values are no longer degenerate.

Problems

5.1 Calculate the reduced mass, μ, of an electron in hydrogen-like atoms with nuclear masses corresponding to:
(a) one proton (H),
(b) 92 protons and 146 neutrons, $^{238}_{92}$U, and
(c) infinity.

5.2 State: (a) the value of the principal quantum number, n,
(b) the value of the orbital angular momentum quantum number, l, for the following orbitals:

1s, 3d, 2p, 4f, 5d, 6f, 5g, 3p, 6d and 4s.

How many of each type of orbital may exist in any one atom?

5.3 State which of the following types of orbital are allowed by the quantum rules (Section 5.1):

$1s, 2d, 1p, 3p, 4d, 4i, 3f, 5p, 7g, 2f.$

5.4 Use equation (5.18) to calculate the ionization potentials of the hydrogen-like atoms, He^+, Li^{2+}, Be^{3+} and B^{4+}. Compare the answers with the experimentally derived values which may be obtained from Appendix II.

5.5 Look again at problems (2.3) and (2.4) in the light of understanding derived from Chapter 5.

5.6 The first ionization potential of sodium is two-fifths that of hydrogen. Calculate the effective nuclear charge of the sodium atom as far as the $3s$ electron is concerned.

6

The Periodic Classification of the Elements

6.1 Electron Spin and Atomic Emission Spectroscopy

In the very familiar flame test for sodium a characteristic yellow emission is produced when sodium chloride is introduced into a gas–air flame. When analyzed using a monochromator of high resolving power this characteristic sodium emission is observed to be a doublet. There are two emission lines with wavelengths 588·9963 nm and 580·5930 nm. Many other emission lines are found to be doublets upon closer inspection and other multiplets (triplets, quartets, etc.) are also observed in the spectra of various elements. The explanation of the doublet lines came from the results of experiments carried out by Stern and Gerlach (1921) who found that when a beam of atoms was allowed to pass through an inhomogeneous magnetic field, the beam was split into several separate beams. When the experiment was carried out with atoms of silver the beam was split into two. The interpretation was that atoms of silver could possess one of two possible orientations of a magnetic moment. The same result was obtained with sodium and potassium atoms. According to classical theory the magnetic moments of the atoms should have been capable of all possible orientations in the magnetic field and the beam should have been spread out evenly. However, the two possible orientations observed in an applied magnetic field indicates quantization of such orientation.

Uhlenbeck and Goudsmit observed the doublet and other multiplet splittings of atomic emission lines and they suggested an explanation in terms of the assignment of *spin angular momentum* to the electron. That is, as well as *orbital angular momentum* an electron possesses *spin angular momentum*, the two momenta interacting in a quantized manner.

The orbital angular momentum is related to the value of the quantum number, l, and a new quantum number, s, denotes the spin angular momentum. The latter quantum number may have one of two values, these being $+\frac{1}{2}$ or $-\frac{1}{2}$ (or sometimes α and β), the former having the values previously described in Section 5.1.

Although the modes of movement of electrons in orbitals are uncertain, it is useful to realize that such motion as there is of the electron will produce an associated magnetic field. In this way the assignment of orbital and spin magnetic moments to the moving electron may be understood. In the absence of an applied magnetic field an electron will possess both orbital and spin angular momenta which are defined by the values of l and s respectively. The resultant angular momentum of the electron is dependent upon the quantized method of the interaction between the orbital and spin angular momenta. This interaction may be dealt with in terms of the interaction between the two quantum numbers, l and s. In general the interaction can be written as:

$$l+s, l+s-1, l+s-2, \ldots, l-s$$

the resulting quantum number being designated by j, with the provision that j must not be negative since it represents resultant angular momentum.

When more than one electron is to be considered it is normal to consider (a) the resultant orbital angular momentum, (b) the resultant spin angular momentum of an atom, and finally (c) the interaction between the resultant orbital and spin angular momenta in describing the *state* of the atom.

The various electrons to be considered may have individual l values, $l_1, l_2, l_3, \ldots$, the resultant of their interaction being L.

Likewise, the individual values of s may be $s_1, s_2, s_3, \ldots$, their resultant being S. Finally the resultant angular momentum of the atom depends on the interaction between L and S to give J.

Two rather simple examples will be treated here.

(1) *The sodium atom* has a single valency electron which happens to reside normally in a $3s$ orbital. All the other electrons are paired up and may be ignored since their orbital and spin momenta cancel out. So, in the ground state of the sodium atom, $l = 0$ and $s = \frac{1}{2}$ for the only electron to be considered. The j value is $\frac{1}{2}$.

If the $3s$ electron is promoted to one of the $3p$ orbitals then l is now 1 and s remains as $\frac{1}{2}$. There are two different j values permitted by the quantum rules. These are $l+s = \frac{3}{2}$ and $l-s = \frac{1}{2}$. Both situations may arise with an assembly of electronically excited sodium atoms and represent levels of different total energy.

(2) *The mercury atom* has in its ground state two valency electrons

The Periodic Classification of the Elements

which are paired up in the 6s orbital. For this pair $l_1 = l_2 = 0$ so that $L = 0$, and $s_1 = \frac{1}{2}$, $s_2 = -\frac{1}{2}$ so that $S = 0$ and in consequence $J = 0$. If one of the 6s electrons is promoted to one of the 6p orbitals then several situations may arise. Considering orbital angular momentum we have $l_1 = 0$ and $l_2 = 1$ so that $L = 1$.

The value of the resultant spin quantum number, S, depends upon whether the spin of the promoted electron is conserved during the promotion or whether it has been 'flipped'. With conservation of spin we have $s_1 = \frac{1}{2}$, $s_2 = -\frac{1}{2}$ and $S = 0$, but if the spin of the excited electron is reversed (this is possible now that the two electrons no longer occupy the same orbital) we have $s_1 = \frac{1}{2}$, $s_2 = \frac{1}{2}$ so that $S = 1$.

With spin conservation there is only one resultant J value for the interaction of L and S, this being 1. In the flipped spin case where both electrons possess the same value of the spin quantum number, there are three possible resultant J values, these being given by

$$L+S = 1+1 = 2,$$

$$L+S-1 = 1+1-1 = 1, \quad \text{and}$$

$$L-S = 1-1 = 0.$$

In the case of the mercury atom, then, there are four different levels, each possessing different total energies.

In order to represent the preceding electronic states of atoms in a concise manner a system of coding has been adopted for the L values which is similar to that used for the values of the orbital angular momentum quantum number, l, in that the letters are the same. The only difference is that upper case letters are employed and are called *term* symbols. Some examples are given in Table 6.1. Whether the electrons concerned

Table 6.1. Coding for Resultant Orbital Angular Momentum

L	Code letter for electronic state (term symbol)
0	S
1	P
2	D
3	F
4	G
5	H

in determining the resultant orbital and spin angular momenta of levels are paired or otherwise (and the extent of pairing in the case of several electrons) is of great importance as far as the total energies of the levels are concerned. With such consideration in mind an indication is given of the extent of electron pairing by the *multiplicity* of the state. The multiplicity is simply calculated as being $2S+1$. This number is placed at the top left-hand corner of the term symbol (i.e. the code letter for L) and at the bottom right-hand corner is placed the relevant J value.

Applying these considerations to the two examples dealt with above we conclude that the ground state of the sodium atom is a $^2S_{1/2}$ state, while that of the mercury atom is a $^1S_0$ state. These are known respectively as doublet-S-one-half and singlet-S-nought states.

All singlet states are those in which all the electrons are paired up and represent states of single (and non-degenerate) energy. Doublet states arise from single unpaired electrons and since there are two ways of combining the orbital and spin contributions to the total angular momentum there are usually two different energy states associated with two different J values. Where either L or S are zero this is not possible, the two different combinations resulting in the same total energy. In other words the two states are degenerate. This is the case for the sodium ground state which is nevertheless termed a doublet.

The excited state of the sodium atom with an electron in a $6p$ orbital

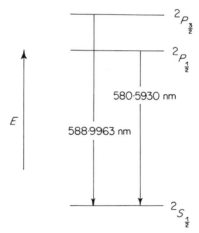

Figure 6.1. The ground and lowest electronically excited states of the sodium atom showing the emissive transitions

corresponds to a ^2P state, and this is a doublet involving two different values of J, the two levels having slightly different energies and are described as $^2P_{3/2}$ and $^2P_{1/2}$ levels respectively. The energy relationships between these and the ground state are shown in Figure 6.1 which indicates the two transitions responsible for the yellow doublet emission which results when the 6p electrons in an assembly of sodium atoms return, radiatively, to the ground state.

The case of the mercury atom is somewhat more complicated than that of the sodium atom. There are states with different multiplicities to consider. First we may deal with the singlet states. The ground state, $^1S_0$, has already been mentioned. The excited state for which there is spin conservation is also a singlet state ($L = 1, S = 0, J = 1$) and is labelled as $^1P_1$.

The state involving the flipped spin has $L = 1$ and $S = 1$ and is a triplet-P state, ^3P. When L, S combination is considered (spin–orbital interaction) this resolves into three levels labelled as $^3P_2$, $^2P_1$ and $^3P_0$. The ground state and the four upper levels are shown in the diagram in Figure 6.2, together with *two* emissive transitions which are observed to occur in an electrical discharge through mercury vapour.

It should be noticed that the triplet levels of the mercury atom are lower in energy than the excited singlet state even though all four levels correspond to the same $6s^16p^1$ electronic configuration. In general, states of highest multiplicity are the lowest in energy for any particular electronic configuration, and this important point will be treated in detail in Section 6.3 (Hund's Rules).

The two observed transitions shown in Figure 6.2 for the mercury atom may be written as:

$$^1P_1 \rightarrow {}^1S_0 \text{ (184·9 nm)}$$

and

$$^3P_1 \rightarrow {}^1S_0 \text{ (253·7 nm)}$$

The latter transition, although it occurs, is known as a forbidden transition. The other two possible transitions to the ground state,

$$^3P_2 \rightarrow {}^1S_0$$

and

$$^3P_0 \rightarrow {}^1S_0$$

are also forbidden and are, in fact, not observed in the mercury emission spectrum.

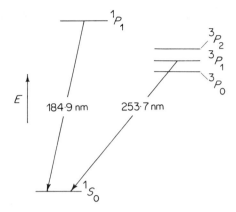

Figure 6.2. The ground and excited states of the mercury atom showing observed emissive transitions

There are two factors operating here that 'forbid' electronic transitions. The first of these may be called *multiplicity or spin forbiddenness* which indicates that for an electronic transition to be allowed the multiplicities of the two states should be identical. If they are not the transition is forbidden. This usually means that the transition probability* is low, which is manifested physically in a low intensity emission (compared to a fully allowed transition as are the two sodium emissions). The $^3P_1 \to {}^1S_0$ is multiplicity forbidden but occurs, and is in fact responsible for a greater intensity of emission at 253·7 nm than the fully allowed $^1P_1 \to {}^1S_0$ transition produces at 184·9 nm. This is due to the transmission properties of the material (silica) from which the discharge lamp is constructed.

The other factor influencing the forbiddenness of electronic transitions is that a selection rule limits allowed transitions to those in which the change in the value of the J quantum number is zero or ± 1 but does not include the $0 \to 0$ case. The $^3P_1 \to {}^1S_0$ transition is thus allowed by this selection rule but the other two triplet-to-ground state transitions are both forbidden.

There is a third restriction of electronic transitions which states that in

---

* The transition probability is, in theory, calculable for any transition. In practice it is only possible to say whether this quantity is finite or zero. If it is zero then the particular transition is said to be forbidden. Even so, some such 'forbidden' transitions do occur at low intensities and exemplify the slight inadequacy of the theoretical treatment.

The Periodic Classification of the Elements

an allowed transition the state symbol must change by one unit. Allowed are:

$$S \to P$$
$$P \to S$$
$$P \to D$$
$$D \to P$$
$$D \to F$$
$$F \to D$$
$$F \to G, \text{etc.}$$

This really means that in a one-electron transition the l value of the electron concerned must change by ± 1 unit. This is the Laporte Rule and allows the mercury transitions dealt with above. It would forbid the $6s^1\, 7s^1 \to 6s^2$ transitions for instance.

It must be emphasized that when electronic transitions are being dealt with the electronic state terms should be considered rather than the actual transition between atomic orbitals. For example, the four possible transitions in the mercury case just discussed are all due to the same orbital change, i.e.

$$6s^1\, 6p^1 \to 6s^2$$

The principles of atomic emission spectroscopy have been dealt with since they have application to the problems arising when atoms combine together to produce molecular systems which will be the subject of the second part of this book.

Effects of Magnetic Fields upon Spectral Lines

The ground state of the calcium atom is $^1S_0$ since the valency electrons ($4s^2$) are paired and in an s orbital. When one of these electrons is promoted to a $4p$ orbital we have to take into account the l and s values of the two electrons in determining the nature and term symbols for the excited state. The term symbol is P since $l_s = 0$ and $l_p = 1$ so that $L = 1$. There are two possibilities for the value of S, these being $S = 1$ ($\frac{1}{2}+\frac{1}{2}$) and $S = 0$ ($\frac{1}{2}-\frac{1}{2}$). The two combinations of s values lead to a triplet state, ^3P, and a singlet state, ^1P.

Let us consider the singlet state in which the corresponding J value ($L+S$) is 1, giving $^1P_1$ as the term symbol.

The electronic transition appearing in the emission spectrum of calcium is shown in Figure 6.3.

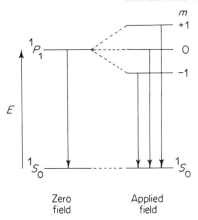

Figure 6.3. The effect of an applied magnetic field upon the $^1P_1 \to {}^1S_0$ transition of the calcium atom

When an external magnetic field is applied to the emitting calcium atoms, it is found that this transition is replaced by a triplet of lines as would be explained if the upper level ($^1P_1$) split into three levels of different energies as shown in Figure 6.3. The quantum theory of electronic levels is such that, for a particular value of J, a level has a degeneracy given by $2J+1$. The ground state for which $J = 0$ is non-degenerate while the excited state ($J = 1$) is triply degenerate. In a magnetic field it is possible for this degeneracy to be broken and thus three lines are produced in the emission spectrum where before there was only one. Thus the quantization of J values in a magnetic field may explain what is called the *Zeeman effect*.

On an orbital basis this may be explained in terms of the quantized alignments of the orbital and spin angular momenta with respect to the direction of an applied magnetic field. The orbital contribution is aligned according to the various possible values of l (in the case of one electron) as given by the permitted values of the magnetic quantum number, m. The spin contribution combines with the various l values (sometimes known as m_l values in the presence of an applied magnetic field) in the usual way. Likewise for a combined L value there are a limited number of ways ($2L+1$) in which the orbital angular momentum vector can be aligned in an applied magnetic field. In the absence of such a field each state is ($2L+1$)fold degenerate, just as a set of orbitals of a particular l value are ($2l+1$)fold degenerate. The quantized alignments of the spin–orbital coupled angular momenta with respect to an applied magnetic field explain the Stern–Gerlach experiments on atomic beams.

The Periodic Classification of the Elements

The above represents the absolute minimum of introduction to the treatment of electronic states and the Zeeman effect, and other books should be consulted for more complicated cases.

6.2 Electron Pairing and Electron Correlation

In Section 2.5, evidence was presented for the existence of electron pairs in atoms. It is now appropriate to deal in more detail with this point.

Previous treatment of the theory of the atom has been restricted to a consideration of atomic orbitals. It is of fundamental importance to understand the effects of actually placing electrons in these orbitals. We have seen that by having more than one electron in an atom the simple theory breaks down and that electron shielding effects have to be considered (Section 5.5). The detailed treatment of interelectronic effects is known as *electron correlation*. The topic of electron correlation may be divided into two parts, these being known as *charge correlation* and *spin correlation* respectively.

Charge correlation is concerned with the fact that electrons are negatively charged and will repel each other according to Coulomb's Law. This means that the further away from each other they are, the more stable is the system containing them. The attraction between the nucleus and the electrons balances this effect to a certain extent, otherwise atoms would not have any stability.

Spin correlation is a more complex concept. Let us consider two orbitals which are described by the wave functions, ψ_a and ψ_b. Also, let us consider the distribution of two electrons in these orbitals. If electron 1 is in orbital a and electron 2 is in orbital b, the state may be described by the wave function,

$$\psi = \psi_a(1) \cdot \psi_b(2) \tag{6.1}$$

The function $\psi_a(1)$ is related to the probability of finding electron 1 in the orbital a, while $\psi_b(2)$ is related to the probability of finding electron 2 in orbital b. ψ as given by equation (6.1) is related to the probability of finding electron 1 in orbital a, *and* of finding electron 2 in orbital b. The individual functions, $\psi_a(1)$ and $\psi_b(2)$, are multiplied since the probability of simultaneous occurrence of two events is given by the product of the probabilities of their separate occurrence.

Alternatively we may have electron 1 in orbital b and electron 2 in orbital a, this state being described by a different wave function ψ'.

$$\psi' = \psi_a(2) \cdot \psi_b(1) \tag{6.2}$$

Now ψ and ψ' are clearly not identical as written but they must represent identical situations since electrons are indistinguishable from each other. In order to build into the system the indistinguishability of electrons, we have to use the theorem which states that since ψ and ψ' are separately acceptable solutions to the wave equation for the total system, then any linear combination is also an acceptable solution (see Section 5.2).

The simplest linear combinations are:

$$\psi_S = \psi + \psi' \tag{6.3}$$

and

$$\psi_{AS} = \psi - \psi' \tag{6.4}$$

In both of these new wave functions the question of electron distinguishability does not arise since all combinations of electrons 1 and 2 appear with the orbitals a and b in ψ_S and ψ_{AS}.

The two wave functions, ψ_S and ψ_{AS}, differ in a very important respect. Consider ψ_S first, this being written more fully as:

$$\psi_S = \psi_a(1) \cdot \psi_b(2) + \psi_a(2) \cdot \psi_b(1) \tag{6.5}$$

If electrons 1 and 2 are interchanged, this becomes

$$\psi_S = \psi_a(2) \cdot \psi_b(1) + \psi_a(1) \cdot \psi_b(2) \tag{6.6}$$

which is identical to equation (6.5). The wave function, ψ_S, is said, therefore, to be *symmetric* with respect to the operation of interchanging the electrons. This is not the case with equation (6.4) which may be written in full as:

$$\psi_{AS} = \psi_a(1) \cdot \psi_b(2) - \psi_a(2) \cdot \psi_b(1) \tag{6.7}$$

and which, upon the interchange of electrons, becomes:

$$\psi'_{AS} = \psi_a(2) \cdot \psi_b(1) - \psi_a(1) \cdot \psi_b(2) \tag{6.8}$$

where

$$\psi'_{AS} = -\psi_{AS} \tag{6.9}$$

i.e. the wave function changes sign with the operation and is said to be *antisymmetric* with respect to it. Both the *symmetric* function, ψ_S, and the *antisymmetric* function, ψ_{AS}, are satisfactory wave functions as far as they express the indistinguishability of electrons.

However, they cannot both be valid functions for a real system at the same time since, if they are, a linear combination of them would be valid and this would lead straight back to equations (6.1) and (6.2) which assert the distinguishability of the particles. The conclusion is that for real

The Periodic Classification of the Elements

systems the total wave function must be symmetric or antisymmetric with respect to electron interchange, *it cannot be both*.

For electrons (but not necessarily for other particles) the total wave function must be antisymmetric. This is a statement which cannot be derived from theory but follows from observation, and is one way of stating the *Pauli Exclusion Principle*. Alternative ways of stating this entirely empirical but tremendously important principle are that:

(1) no two electrons in an atom may have identical wave functions, and
(2) no two electrons in an atom may have identical sets of values of the four quantum numbers, n, l, m and s.

The orbital wave functions which have been dealt with up to now do not represent the total wave functions of electrons. The total wave function must include a spin wave function which we can write in terms of α and β (since there are only two possible wave functions). The total electronic wave function of a system may be written as the product of the orbital and spin functions.

$$\psi_{total} = \psi_{orbital} \cdot \psi_{spin} \quad (6.10)$$

For two electrons, 1 and 2, in the ψ_a and ψ_b orbitals we can write the total wave functions,

$$\psi_1 = [\psi_a(1) \cdot \psi_b(2) + \psi_a(2) \cdot \psi_b(1)] \cdot \psi_{spin} \quad (6.11)$$

and

$$\psi_2 = [\psi_a(1) \cdot \psi_b(2) - \psi_a(2) \cdot \psi_b(1)] \cdot \psi_{spin} \quad (6.12)$$

If the electrons possess identical spins, there would be two symmetric spin functions which would express the situation, these being written as:

$$\alpha(1) \cdot \alpha(2)$$

and

$$\beta(1) \cdot \beta(2)$$

If the two electrons have opposite spins there would again be two spin functions to express the situation, these being written as:

$$\alpha(1) \cdot \beta(2) + \alpha(2) \cdot \beta(1)$$

and

$$\alpha(1) \cdot \beta(2) - \alpha(2) \cdot \beta(1)$$

the former being symmetric, and the latter being antisymmetric, to electron exchange (not $\alpha(1).\beta(2)$ and $\alpha(2).\beta(1)$, since these would assert distinguishability of the two electrons).

To select the appropriate spin functions for equations (6.11) and (6.12), it is necessary to invoke the Pauli Principle which, applied to electrons, means that the total wave functions expressed by equations (6.11) and (6.12) must be antisymmetric to electron interchange.

The possible combinations are of a symmetric orbital function with an antisymmetric spin function or an antisymmetric orbital function with a symmetric spin function. Combination of two symmetric or two antisymmetric functions would violate the Pauli Principle, both resulting in symmetric total wave functions.

In the case being considered the antisymmetric total wave functions are:

$$\psi_1 = [\psi_a(1).\psi_b(2)+\psi_a(2).\psi_b(1)][\alpha(1).\beta(2)-\alpha(2).\beta(1)] \quad (6.13)$$

$$\psi_2 = [\psi_a(1).\psi_b(2)-\psi_a(2).\psi_b(1)][\alpha(1).\alpha(2)] \quad (6.14)$$

$$\psi_3 = [\psi_a(1).\psi_b(2)-\psi_a(2).\psi_b(1)][\beta(1).\beta(2)] \quad (6.15)$$

$$\psi_4 = [\psi_a(1).\psi_b(2)-\psi_a(2).\psi_b(1)][\alpha(1).\beta(2)+\alpha(2).\beta(1)] \quad (6.16)$$

Of these four viable wave functions for a two electron system the first, ψ_1, is a singlet state with the spins of the electrons in opposition. In terms of electronic states $s_1 = \frac{1}{2}$ and $s_2 = -\frac{1}{2}$ so that $S = 0$ and $2S+1 = 1$. It is of tremendous importance to note that if we make $\psi_a = \psi_b$ (i.e. we are dealing with a single orbital) then ψ_1 becomes:

$$\psi'_1 = [\psi_a(1).\psi_a(2)+\psi_a(2).\psi_a(1)][\alpha(1).\beta(2)-\alpha(2).\beta(1)] \quad (6.17)$$

$$= [2\psi_a(1).\psi_a(2)][\alpha(1).\beta(2)-\alpha(2).\beta(1)] \quad (6.18)$$

which is *finite* indicating that it is possible for two electrons to occupy the same orbital *provided* that their spins are in opposition.

The other three functions, ψ_2, ψ_3 and ψ_4, form a triplet state* in which the orbital function is constant. The spins of the electrons are parallel and so $S = 1$ and $2S+1 = 3$, hence the triplet state. Also of interest is that if we now make $\psi_a = \psi_b$, the triplet state functions (6.14, 6.15 and 6.16) all reduce to zero, which means that it is not possible for a triplet state to exist with two electrons in one orbital. There is zero probability of having parallel spins if the electrons are in the same orbital. The total

* In a magnetic field there are three possible orientations of the spin vector, S, given by $S = 1, 0$ and -1. These correspond to the three spin functions used to describe the triplet state.

The Periodic Classification of the Elements

wave functions for the singlet and triplet states can be evaluated reasonably easily if ψ_a and ψ_b are s orbitals. An important conclusion can be drawn from the results of such calculations for the excited helium atom configuration, $1s^1 2s^1$, i.e. helium in which one electron is in the 1s orbital and the other is in the 2s orbital. The radial probabilities of finding simultaneously the first electron at a distance r_1 and the second at a distance r_2 from the nucleus are plotted as contour diagrams for the singlet state in Figure 6.4 and for the triplet state in Figure 6.5. The r

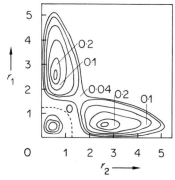

Figure 6.4. Contour diagram of the probability of electron 1 being at r_1 and of electron 2 being at r_2 for the helium atom in its $1s^1 2s^1$ singlet state

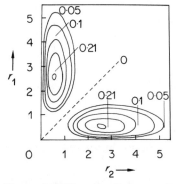

Figure 6.5. Contour diagram of the probability of electron 1 being at r_1 and of electron 2 being at r_2 for the helium atom in its $1s^1 2s^1$ triplet state

(Both of these figures are taken from *Wave Mechanics and Valency*, by J. W. Linnett, Methuen, 1960; reproduced by permission of the author)

values plotted in Figures 6.4 and 6.5 are multiples of the maximum in the radial distribution function of the hydrogen atom (0·053 nm).

Figure 6.4 shows that for the singlet state three spatial arrangements have high probability, these consisting of two in which one electron is near the nucleus and one further away, and the third in which *both electrons* are near the nucleus. Figure 6.5 shows that for the triplet state only two arrangements are of high probability, these being where one electron is close to the nucleus and the other is further away. The dotted lines in both figures indicate regions of zero probability of finding electrons, and for the triplet state show that there is zero probability for $r_1 = r_2$, i.e. no possibility for the electrons to be in the same region of the atom.

It can be calculated for the $1s^1 2s^1$ state of helium that the unperturbed state (i.e. no electron correlation) has an energy of $5 E_H$ where E_H is the energy of an electron in the 1s orbital of the hydrogen atom (remember that E_H is negative). The eigenvalues are given by equation (5.4), and the 1s orbital of hydrogen has the energy,

$$E_H = -\frac{2\pi^2 \mu e^4}{h^2} \tag{6.19}$$

so that equation (5.4) may be rewritten as

$$E = \frac{E_H Z^2}{n^2} \tag{6.20}$$

Thus for the 1s orbital of helium ($Z = 2$) (neglecting the correction of μ in equation 5.4), we have

$$E_{1s} = \frac{E_H 2^2}{1^2} = 4E_H \tag{6.21}$$

and for the 2s orbital

$$E_{2s} = \frac{E_H 2^2}{2^2} = E_H \tag{6.22}$$

The total unperturbed energy of one electron in each of these orbitals being given by

$$E_{1s} + E_{2s} = 5E_H \tag{6.23}$$

Linnett has calculated the effects of:

(1) coulombic interaction of the two electrons, and
(2) spin correlation

The Periodic Classification of the Elements

and by comparison with the experimental energies for the singlet and triplet states has deduced the extent of charge correlation.

These effects are shown in Figure 6.6. The coulombic interaction, the mean interaction between the two electrons, destabilizes the unperturbed state as would be expected since there is repulsion between the two electrons. The spin correlation effect destabilizes the singlet state further but stabilizes the triplet state. Charge correlation produces a stabilization due to modification of the orbitals by repulsion between the electrons, the singlet state being stabilized to a greater extent than the triplet state

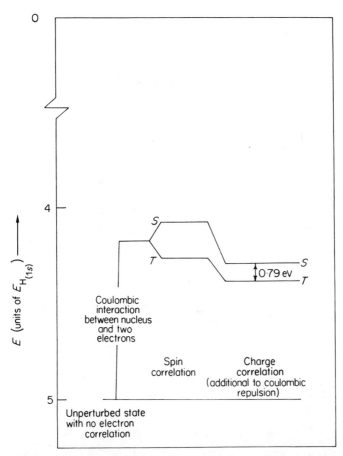

Figure 6.6. A diagram showing the coulombic interaction between two electrons, together with spin and charge correlation effects on the $1s^1 2s^1$ states of the helium atom

since, in the singlet state, spin correlation has brought the two electrons together and, in the triplet state, has forced them apart.

For the ground state of helium where we have both electrons in the 1s orbital, the spin correlation energy is zero since we are considering only one orbital. The coulomb energy for this state is $-2.5\, E_H$ and the charge correlation energy is $0.308\, E_H$ which is even greater than in the $1s^1 2s^1$ singlet state, since the electrons are even closer together in the $1s^2$ state. The 1s orbital is also smaller in volume than the 2s orbital. The charge correlation energy is calculated as the difference between the observed ionization potential of helium and that derived from the

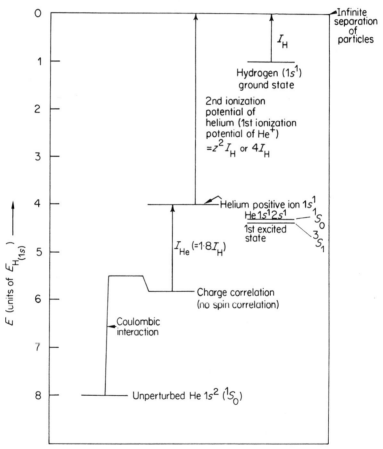

Figure 6.7. A diagram which shows the estimation of charge correlation in the ground state of the helium atom, and compares this with the $1s^1 2s^1$ excited state and the energies of He^+ and H

The Periodic Classification of the Elements

unperturbed energy and the coulombic interaction. This is shown in Figure 6.7, together with the energies of H ($1s^1$), He$^+$ ($1s^1$) and He ($1s^1\ 2s^1$), and their observed ionization potentials.

The consequences of electron correlation may be summarized as follows:

(1) *Spin correlation* tends to make electrons of opposed spins come together, and electrons of the same spins (parallel) keep apart.
(2) *Charge correlation* in addition to the coulombic repulsion occurring between electrons involves a stabilization of electrons due to the *modification* of the orbitals they accommodate by electronic repulsion. The charge correlation effect offsets the coulombic repulsion to a small extent and is greater in singlet states than in triplet states since in the former the electrons are closer together due to spin correlation.

6.3 The Pauli Exclusion Principle and the Construction of the Periodic Classification of the Elements

We have seen that the Pauli Principle may be stated in various ways, the one which is best applied to atom building being that which states that no two electrons in an atom may have identical sets of values for the four quantum numbers. If we consider an orbital with an orbital function, ψ, containing two electrons, then the two electrons have identical values of the three quantum numbers, n, l and m, and the orbital wave function may be written as:

$$\psi = \psi_a(1) \cdot \psi_a(2) \qquad (6.24)$$

which is symmetric to electron exchange. For the total wave function to be antisymmetric to electron exchange the spin function must be antisymmetric, i.e.

$$\alpha(1) \cdot \beta(2) - \alpha(2) \cdot \beta(1)$$

indicating that the two electrons must have opposite values of the spin quantum number.

The consequences of the Pauli Principle are that the maximum number of electrons allowed to occupy an atomic (or any other) orbital is *two* and that only *one* orbital for a given set of n, l and m values is allowed for any one atom. If we know the order of energies (eigenvalues) of these orbitals we can then build up the periodic classification of the elements. The order of filling of atomic orbitals has been determined experimentally and the order of energies has been shown in Figures 5.11 and 5.12.

Let us now consider the application of the preceding principles and experimental facts to the construction of the electronic configuration of the elements. The working principle is to decide first of all on the atomic orbitals available, their number and finally their order of energies, starting with the bare nucleus to feed in Z electrons (where Z is the nuclear charge). The constructional procedure is sometimes referred to as the *aufbau principle*.

First, then, we have to decide upon the available orbitals, bearing in mind the quantum number rules (Section 5.1). The orbitals are

$1s, 2s, 2p, 3s, 3p, 3d, 4s, 4p, 4d, 4f, 5s, 5p, 5d, 5f, 5g$, etc.

The Pauli Principle limits us to one $1s$, one $2s$, three $2p$, one $3s$, three $3p$, five $3d$ orbitals, etc., per atom. The order of energies for a given value of Z is known (Figure 5.12). We can now build some atoms. Starting with the hydrogen atom we have one electron to place. This will occupy the lowest vacant atomic orbital which is the $1s$. The electronic configuration of hydrogen in its ground state is therefore

$$\text{H } 1s^1 \quad \text{or} \quad \boxed{\uparrow\ }$$

With helium ($Z = 2$) the first electron will give a hydrogen-like configuration to the helium positive ion:

$$\text{He}^+ \, 1s^1 \quad \text{or} \quad \boxed{\uparrow\ }$$

and the second electron, overcoming coulombic repulsion will also go into the $1s$ orbital pairing up with the first electron:

$$\text{He } 1s^2 \quad \text{or} \quad \boxed{\uparrow\downarrow}$$

The excited configuration, $1s^1 2s^1$, which has been mentioned already (Section 6.2, Figures 6.4 and 6.5) is less stable even though a triplet (spin correlation stabilized) configuration is possible. This is because the energy gap between the $1s$ and $2s$ orbitals in helium ($3E_\text{H}$) is greater than the coulombic repulsion energy of pairing in the $1s^2$ state ($2 \cdot 5 \, E_\text{H}$).

In the case of lithium ($Z = 3$) the first two electrons occupy the $1s$ orbital giving the helium configuration $1s^2$ to the Li^+ ion, and the third electron occupies the lowest vacant orbital which is the $2s$, giving the configuration for lithium as $1s^2 2s^1$ which is shown diagrammatically in Figure 6.8.

The Periodic Classification of the Elements

Figure 6.8. The electronic configuration of the lithium atom

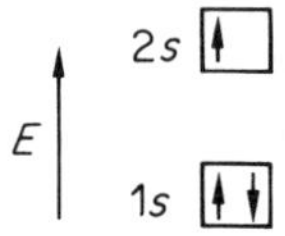

Figure 6.9. The electronic configuration of the boron atom

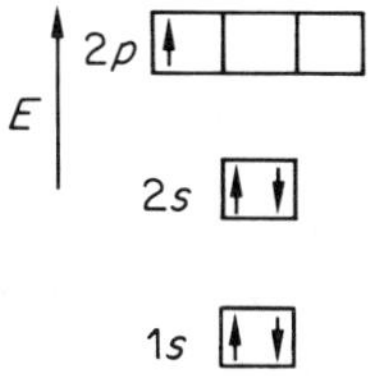

The fourth electron in beryllium ($Z = 4$) fills the 2s orbital (rather than simply occupying the higher energy 2p orbital) to give the configuration for beryllium as:

$$1s^2\, 2s^2 \quad \text{or} \quad (\text{He})\, 2s^2$$

In boron ($Z = 5$) the fifth electron occupies one of the three degenerate 2p orbitals. It is usual to consider that the electron goes into the $2p_x$ orbital, but due to the degeneracy of the set of three 2p orbitals, this is completely arbitrary. The configuration for boron is

$$1s^2\, 2s^2\, 2p_x^1 \quad \text{or} \quad (\text{He})\, 2s^2\, 2p_x^1$$

which is shown in Figure 6.9.

We come now to a very interesting situation, for in the placing of the sixth electron in the electronic structure of carbon ($Z = 6$) we have a choice between pairing the electron with the one already occupying the $2p_x$ orbital to give the configuration, $(\text{He})\, 2s^2\, 2p_x^2$, or to place the sixth electron in an otherwise vacant orbital, say the $2p_y$ (chosen arbitrarily) to give the configuration, $(\text{He})\, 2s^2\, 2p_x^1\, 2p_y^1$, which can be either a singlet or triplet state. This is a general point concerning the filling of a set of *degenerate* orbitals and is very important in the determination of the ground state configurations of atoms and molecules. Bearing in mind the results of electron correlation it is possible to predict the correct ground state. Of the three possibilities mentioned, two are singlet states and the other is a triplet state. Since all three states are concerned with electron distribution within the same degenerate set of orbitals we can say that the triplet state will be the lowest in energy since spin correlation stabilization will be greatest for this state. Of the two singlet states the former, $(He) 2s^2 2p_x^2$, will be the least stable because of the lack of any spin correlation stabilization and a large coulombic repulsion destabilization. Two generalizations are possible arising from these considerations which always apply to the addition of electrons to degenerate levels. The first is that the number of unpaired electrons or the number of singly occupied orbitals will be a maximum (coulombic effect), and secondly these electrons will have parallel spins, i.e. the same value of the spin quantum number (spin correlation effect). These two generalizations are a common form of what are known as Hund's Rules, more examples of which will be dealt with below.

Continuing with nitrogen ($Z = 7$), the seventh electron by the above consideration will minimize coulombic repulsion by going into the $2p_z$ orbital and will maximize the spin correlation stabilization by having a spin parallel to those of the electrons already singly occupying the $2p_x$ and $2p_y$ orbitals, so that the configuration may be written as:

$$(He) 2s^2 2p_x^1 2p_y^1 2p_z^1$$

With oxygen ($Z = 8$) the first seven electrons will give the nitrogen configuration, the eighth electron pairing up in the $2p_x$ orbital (or the entirely equivalent $2p_y$ or $2p_z$ orbitals) to give the configuration,

$$(He) 2s^2 2p_x^2 2p_y^1 2p_z^1$$

In a similar way the electronic configuration of fluorine ($Z = 9$) and neon ($Z = 10$) are:

$$(He) 2s^2 2p_x^2 2p_y^2 2p_z^1 \quad \text{and} \quad (He) 2s^2 2p_x^2 2p_y^2 2p_z^2$$

respectively, the orbitals with $n = 2$ being completely filled in neon. The eight elements lithium to neon form what is called the first short period of the periodic classification of the elements. Although the periodic classification was almost complete in its modern form before our knowledge of electronic configuration was as advanced as it is at the present time, it is now possible to understand its structure in a more fundamental manner. The periods are so arranged as to form vertical groups of elements with electronic configurations which are almost identical, the only differ-

The Periodic Classification of the Elements

ence being that the value of the principal quantum number of the electron level of highest energy varies down any group. (This statement is qualified below.)

The next element in the periodic system based upon Z values is sodium ($Z = 11$) where ten electrons assume the neon configuration denoted by (Ne), the eleventh electron occupying the next vacant level which is the $3s$ orbital. The configuration of sodium is therefore (Ne) $3s^1$. This is the first element in the second short period of eight elements, having this place historically because of its chemical similarity to lithium and, on an electronic basis, because both elements have a so-called inert gas configuration plus a singly occupied s orbital. The latter differs only in the value of n. Apart from this difference, the electronic configurations of the other seven elements in the second short period are entirely analogous to those of the elements beryllium to neon and have the configurations:

Magnesium	Mg	(Ne)	$3s^2$
Aluminium	Al	(Ne)	$3s^2\,3p_x^1$
Silicon	Si	(Ne)	$3s^2\,3p_x^1\,3p_y^1$
Phosphorus	P	(Ne)	$3s^2\,3p_x^1\,3p_y^1\,3p_z^1$
Sulphur	S	(Ne)	$3s^2\,3p_x^2\,3p_y^1\,3p_z^1$
Chlorine	Cl	(Ne)	$3s^2\,3p_x^2\,3p_y^2\,3p_z^1$
Argon	Ar	(Ne)	$3s^2\,3p_x^2\,3p_y^2\,3p_z^2$

The next stage in the construction of the periodic table is to assume an argon 'core' for all the elements in the next period which starts with potassium ($Z = 19$) with the nineteenth electron occupying the $4s$ orbital, the configuration being written as (Ar) $4s^1$. The $4s$ orbital which comes below the $3d$ orbitals due to orbital penetration effects (Section 5.5) is filled in calcium ($Z = 20$) whose configuration is (Ar) $4s^2$. The twenty-first electron in scandium ($Z = 21$) occupies the $3d$ level and since there is no analogous configuration in the second short period, this element is the first member of the 'transition' element groups and is the reason for the gap between the *main* element groups two and three. The transition elements may be defined as those in which the d levels are being filled, or alternatively as those elements with an incomplete d set of electrons. The latter definition is more in keeping with those elements which have typical transition element properties, the former definition including those elements where the d levels are completely filled, and which possess properties which are more typical of the main group elements.

After scandium there is a regular filling of the $3d$ levels in accordance with Hund's Rules so we have

Titanium   Ti (Ar) $4s^2 3d^2$

Vanadium   V   (Ar) $4s^2 3d^3$

The ground state configuration of chromium ($Z = 24$) is $(Ar)\,4s^1\,3d^5$ and not $(Ar)\,4s^2\,3d^4$ as might have been expected for a regular filling of the $3d$ orbitals.

At this point ($Z = 24$) the energies of the $4s$ and $3d$ levels are very similar and the extra spin correlation stabilization achieved in the $4s^1\,3d^5$ configuration over the $4s^2\,3d^4$ tips the balance in favour of the former structure. A point to note is that with the $3d^5$ configuration the $3d$ orbitals are exactly half-filled and the spin correlation stabilization is at its maximum.

In manganese ($Z = 25$) the $3d^5$ arrangement is retained, the electronic structure being $(Ar)\,4s^2\,3d^5$ and, after this, a regular filling of the singly occupied $3d$ orbitals ensues giving the configurations:

Iron   Fe (Ar) $4s^2 3d^6$

Cobalt   Co (Ar) $4s^2 3d^7$

Nickel   Ni (Ar) $4s^2 3d^8$

With copper ($Z = 29$) there is another irregularity, the configuration being

$$(Ar)\,4s^1\,3d^{10}$$

This has consequences which bear upon the definitions of transition elements already referred to for, in the chemistry of copper in its $+1$ oxidation state, we have an electronic configuration:

$$Cu^+ \;(Ar)\,3d^{10}$$

which has a full $3d$ shell and does not exhibit typical transition element properties, whereas the $+2$ oxidation state,

$$Cu^{2+}\;(Ar)\,3d^9$$

does not have this restriction and has properties which are typical of species with an incomplete d shell.

These typical properties may be summarized:

(1) paramagnetism due to unpaired d electrons, and
(2) absorption in the visible region producing coloured compounds.

The Periodic Classification of the Elements 103

Copper (I) compounds are usually white (no absorption in the visible region) and are diamagnetic (zero paramagnetic moment), while copper (II) compounds are usually blue or green and have a paramagnetic moment due to the single unpaired $3d$ electron.

In the final transition group (using the first definition of a transition element) we have in zinc ($Z = 30$) filled both the $3d$ and $4s$ orbitals, giving the configuration,

$$(Ar)\, 4s^2\, 3d^{10}$$

Zinc forms only a $+2$ oxidation state which shows no typical transition element properties.

The next orbitals to be used are the $4p$ set which are filled in a regular manner from gallium to krypton, the configurations being

Gallium	Ga	$(Ar)\, 4s^2\, 3d^{10}\, 4p^1$
Germanium	Ge	$(Ar)\, 4s^2\, 3d^{10}\, 4p^2$
Arsenic	As	$(Ar)\, 4s^2\, 3d^{10}\, 4p^3$
Selenium	Se	$(Ar)\, 4s^2\, 3d^{10}\, 4p^4$
Bromine	Br	$(Ar)\, 4s^2\, 3d^{10}\, 4p^5$
Krypton	Kr	$(Ar)\, 4s^2\, 3d^{10}\, 4p^6$

At this point the first long period is complete and with rubidium ($Z = 37$) the thirty-seventh electron occupies the $5s$ orbital, this being lower in energy than the $4d$ and $4f$ levels, giving the configuration, $(Kr)\, 5s^1$.

The $5s$ orbital is filled in the case of strontium ($Z = 38$), $(Kr)\, 5s^2$, and in yttrium ($Z = 39$). The last electron of yttrium occupies one of the $4d$ orbitals and the element is the first member of the second transition series, having the configuration, $(Kr)\, 5s^2\, 4d^1$.

The filling of the $4d$ levels continues with zirconium ($Z = 40$), $(Kr)\, 5s^2\, 4d^2$. In niobium ($Z = 41$) the $(Kr)\, 5s^2\, 4d^3$ configuration is discarded in favour of $(Kr)\, 5s^1\, 4d^4$, presumably due to the greater spin correlation stabilization energy of the state of higher multiplicity (since the $5s$ and $4d$ levels are extremely similar in energy for niobium and the next four elements). We continue therefore with the configurations:

Molybdenum	Mo	$(Kr)\, 5s^1\, 4d^5$
Technetium	Tc	$(Kr)\, 5s^1\, 4d^6$
Ruthenium	Ru	$(Kr)\, 5s^1\, 4d^7$
Rhodium	Rh	$(Kr)\, 5s^1\, 4d^8$

at which point the 4d level becomes distinctly lower in energy than the 5s and is used exclusively in palladium ($Z = 46$), (Kr) $4d^{10}$. The 5s orbital is once again used in silver ($Z = 47$), (Kr) $5s^1 4d^{10}$, and in cadmium ($Z = 48$), (Kr) $5s^2 4d^{10}$. This completes the second transition series, and the building-up process continues with the regular filling of the 5p orbitals in a manner which is consistent with Hund's Rules, with the configurations:

Indium	In	(Kr) $5s^2 4d^{10} 5p^1$
Tin	Sn	(Kr) $5s^2 4d^{10} 5p^2$
Antimony	Sb	(Kr) $5s^2 4d^{10} 5p^3$
Tellurium	Te	(Kr) $5s^2 4d^{10} 5p^4$
Iodine	I	(Kr) $5s^2 4d^{10} 5p^5$
Xenon	Xe	(Kr) $5s^2 4d^{10} 5p^6$

Xenon is the final element in the second long period. The next element is the first in the third long period which is more complex than the other two long periods, since it involves the hitherto unused 4f orbitals. The third long period begins with caesium ($Z = 55$) which, in addition to the xenon 'core', has an electron in a 6s orbital, the configuration being (Xe) $6s^1$.

Barium ($Z = 56$) has a pair of electrons in the same orbital, its configuration being (Xe) $6s^2$, and with lanthanum ($Z = 57$) the third transition series begins with the last electron occupying a 5d orbital, (Xe) $6s^2 5d^1$.

At this point the 5d and 4f levels are very close in energy and instead of continuing with the third transition series there follows the series of fourteen elements known as the Rare Earths or Lanthanides, the latter name coming from the chemical similarities between the elements and lanthanum. The lanthanides are sometimes referred to as the inner transition elements, this description stemming from the fact that the atoms which have 6s and sometimes 5d orbitals occupied have electronic structures which involve the filling of the inner 4f levels. The filling is fairly regular and in cerium ($Z = 58$) the 4f level is definitely below the 5d, giving:

Cerium	Ce	(Xe) $6s^2 4f^2$

This is followed by:

Praesodymium	Pr	(Xe) $6s^2 4f^3$
Neodymium	Nd	(Xe) $6s^2 4f^4$

The Periodic Classification of the Elements

Promethium	Pm	(Xe) $6s^2\ 4f^5$
Samarium	Sm	(Xe) $6s^2\ 4f^6$
Europium	Eu	(Xe) $6s^2\ 4f^7$

at which point the seven $4f$ orbitals will, in accordance with Hund's Rules, be singly occupied and the configuration possesses maximum spin correlation stabilization. For this reason the next electron in gadolinium ($Z = 64$) occupies preferentially a $5d$ orbital, giving

$$(\text{Xe})\ 6s^2\ 5d^1\ 4f^7$$

In terbium ($Z = 65$) the energy gap between the $5d$ and $4f$ levels is sufficient to eliminate the use of the former orbitals, and we have

Terbium	Tb	(Xe) $6s^2\ 4f^9$

From dysprosium ($Z = 66$) to ytterbium ($Z = 70$) there is a regular filling of the $4f$ orbitals giving the configurations:

Dysprosium	Dy	(Xe) $6s^2\ 4f^{10}$
Holmium	Ho	(Xe) $6s^2\ 4f^{11}$
Erbium	Er	(Xe) $6s^2\ 4f^{12}$
Thulium	Tm	(Xe) $6s^2\ 4f^{13}$
Ytterbium	Yb	(Xe) $6s^2\ 4f^{14}$

The fourteenth in the series, lutecium ($Z = 71$) has the configuration,

$$(\text{Xe})\ 6s^2\ 5d^1\ 4f^{14}$$

in which the final electron occupies a $5d$ orbital.

The $5d$ levels are then filled in the elements forming the rest of the third transition series which come after lanthanum. Their configurations are:

Hafnium	Hf	(Xe) $6s^2\ 5d^2\ 4f^{14}$
Tantalum	Ta	(Xe) $6s^2\ 5d^3\ 4f^{14}$
Tungsten	W	(Xe) $6s^2\ 5d^4\ 4f^{14}$
Rhenium	Re	(Xe) $6s^2\ 5d^5\ 4f^{14}$
Osmium	Os	(Xe) $6s^2\ 5d^6\ 4f^{14}$

With iridium ($Z = 77$) there is an irregularity due to the $5d$ level becoming more stable than the 6s, and we have the configuration, (Xe) $5d^9 4f^{14}$.

In platinum ($Z = 78$) spin correlation stabilizes the configuration,

$$(Xe) 6s^1 5d^9 4f^{14}$$

and in gold ($Z = 79$) the filling of the 5d levels is completed with

$$(Xe) 6s^1 5d^{10} 4f^{14}$$

The 6s level is fully occupied in mercury ($Z = 80$) with (Xe)$6s^2 5d^{10} 4f^{14}$, and there follows the regular filling of the 6p orbitals in the elements from thallium to radon, the configurations being:

Thallium	Tl	(Xe) $6s^2 5d^{10} 4f^{14} 5p^1$
Lead	Pb	(Xe) $6s^2 5d^{10} 4f^{14} 5p^2$
Bismuth	Bi	(Xe) $6s^2 5d^{10} 4f^{14} 5p^3$
Polonium	Po	(Xe) $6s^2 5d^{10} 4f^{14} 5p^4$
Astatine	At	(Xe) $6s^2 5d^{10} 4f^{14} 5p^5$
Radon	Rn	(Xe) $6s^2 5d^{10} 4f^{14} 5p^6$

The third long period ends with radon ($Z = 86$) which is the next inert gas, and the fourth long period (which is incomplete) begins with francium ($Z = 87$), having an electron in the 7s orbital in addition to possessing the radon 'core', (Rn) $7s^1$.

Radium ($Z = 88$) has a pair of electrons in the same orbital, (Rn) $7s^2$, and with actinium ($Z = 89$) the fourth transition series starts with the final electron occupying a 6d orbital:

$$Ac \quad (Rn) 7s^2 6d^1$$

By analogy to the rare earth series which was concerned with the filling of the 4f levels, we now expect there to be another series in which the 5f levels are filled. This is so, but the series which are known as actinides starts rather irregularly with thorium ($Z = 90$), having the configuration, (Rn) $7s^2 6d^2$.

It may seem odd that thorium is placed at the beginning of the actinide series instead of being the second member of the fourth transition series, but its chemical properties are more consistent with its actual position.

With protactinium ($Z = 91$) there is still an electron in a 6d orbital and two electrons occupy the 5f levels, giving

$$(Rn) 7s^2 6d^1 5f^2$$

The Periodic Classification of the Elements

the 6d and 5f levels being very close together in energy. In uranium ($Z = 92$) there is retention of the single $6d^1$ configuration, the extra electron going into the 5f level, giving

$$(Rn) \; 7s^2 \; 6d^1 \; 5f^3$$

In neptunium ($Z = 93$) the 5f level is definitely below the 6d and we have the configuration:

        Neptunium   Np $(Rn) \; 7s^2 \; 5f^5$

followed by:

        Plutonium   Pu $(Rn) \; 7s^2 \; 5f^6$

        Americium   Am $(Rn) \; 7s^2 \; 5f^7$

at which point the 5f orbitals are all singly occupied and, as with gadolinium, the f^7 configuration is retained by the next element, curium ($Z = 96$), the extra electron occupying a 6d orbital, giving

        Cm $(Rn) \; 7s^2 \; 6d^1 \; 5f^7$

The next elements have structures which are analogous to the corresponding lanthanides with the configurations:

        Berkelium   Bk $(Rn) \; 7s^2 \; 5f^9$

        Californium   Cf $(Rn) \; 7s^2 \; 5f^{10}$

        Einsteinium   Es $(Rn) \; 7s^2 \; 5f^{11}$

        Fermium   Fm $(Rn) \; 7s^2 \; 5f^{12}$

        Mendelevium   Mv $(Rn) \; 7s^2 \; 5f^{13}$

        Nobelium   No $(Rn) \; 7s^2 \; 5f^{14}$

        Lawrencium   Lw $(Rn) \; 7s^2 \; 6d^1 \; 5f^{14}$

Lawrencium ($Z = 103$) is the last member of the actinide series and is the latest member of the fourth long period. The next elements, if they are even synthesized, would be expected to be members of the fourth transition series involving the filling of the 6d orbitals.

It should be noticed that the ground state electronic structure of the elements eventually use the 7s orbital, the orbitals of lower principal quantum number values which are not used being 5g, 6f, 6g and 6i which, due to electron penetration effects, are all at greater energies than the 7s level.

6.4 Summary

The periodic classification of the elements may be understood in terms of:

(1) the quantum number rules which define the available atomic orbitals (Section 5.1);

(2) the Pauli Exclusion Principle which limits the number of atomic orbitals available and also limits the occupancy of each atomic orbital to a maximum of two electrons (Sections 6.2 and 3); and

(3) the order of filling of the available atomic orbitals which, to a certain extent, may be predicted by theory but when the principal quantum number is greater than 3, it is an experimental order which is used.

When these factors have been taken into account together with Hund's Rules (electron correlation), it is seen that in general each group in the periodic table contains elements which have similar electron configurations. The long form of the periodic table is shown in Figure 6.10 and includes the ground state electronic configurations of the atoms.

Problems

6.1 Calculate the difference in energy between the $^2P_{3/2}$ and $^2P_{1/2}$ states of the sodium atom. Compare this energy with those of the electronic transitions to the $^2S_{1/2}$ state.

6.2 Using Hund's Rules write down in full the ground electronic configurations for the atoms of the following elements:
(a) nitrogen
(b) oxygen
(c) vanadium
(d) iron
(e) europium.

6.3 The information and understanding derivable from this chapter is sufficient to allow a proper theoretical explanation for the ionization potential variations plotted in problems (2.1) and (2.2) and those dealt with empirically in Section 2.5. Problems (2.1) and (2.2) and Section 2.5 should now be considered again.

6.4 The first ionization of an atom takes place from the orbital of highest energy. Note the levels from which ionization occurs for the first eighteen elements on Figure 5.12. Make a separate plot of these orbital levels and compare your results with the first part of the plot in your answer to problem (2.1). Interpret the diagram you obtain in terms of the orbitals involved in the ionization processes.

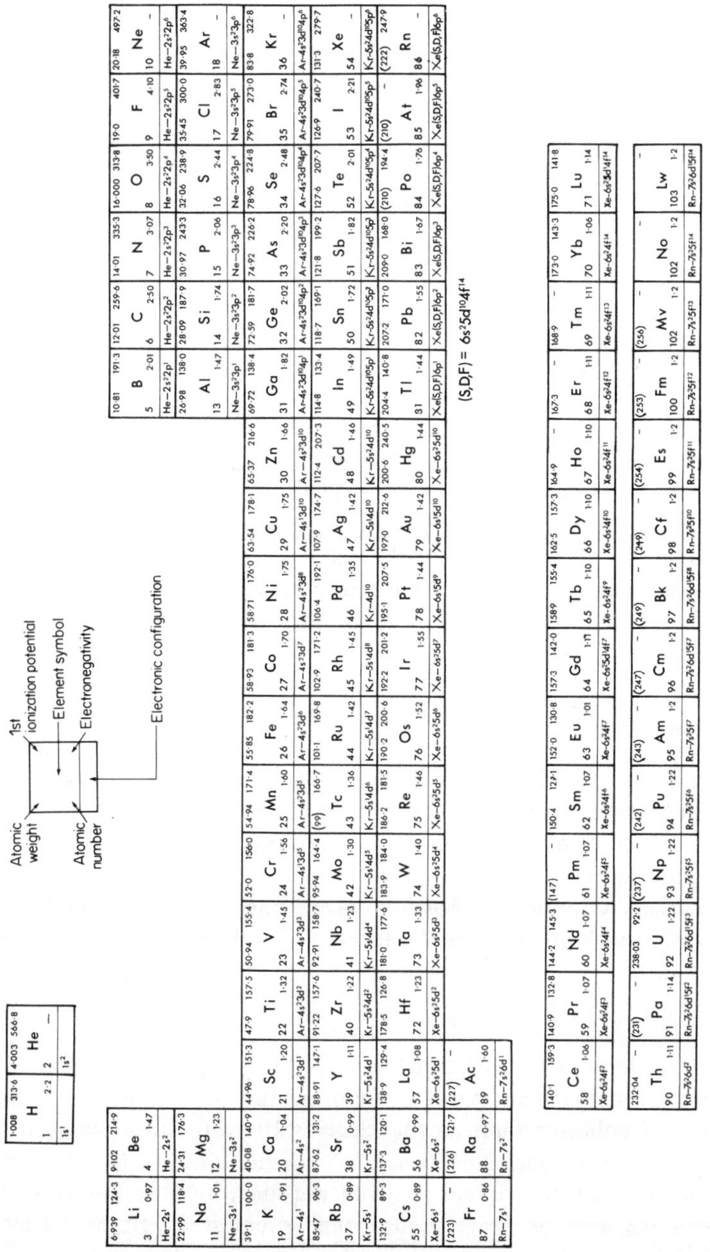

The Periodic Classification of the Elements

Figure 6.10. The periodic classification of the elements

7

Combination of Atoms: The Formation of Diatomic Molecules

7.1 Introduction

When considering the combination of atoms to form molecular systems the wave-mechanical treatments become increasingly complex, and various approximate approaches are used. No details of these will be given but the results of the application of some of these approximate methods will be described. With the wave mechanics so far outlined, a broad level of understanding of molecular systems may be achieved and form a basis for further refined and sophisticated treatments.

7.2 Chemical Bonds

When atoms combine together the product is a molecule. A molecular system is formed when such cohesive forces as are present operate to produce a system which is more stable than the atomic system.

The bonding in molecular systems can be dealt with in terms of two extreme cases. One is *ionic bonding* where one (or more) electron(s) is (are) transferred from one atom to another with the formation of charged atoms or *ions*. The formation of ions may be exemplified by the reaction,

$$Na + Cl \rightarrow Na^+ + Cl^- \tag{7.1}$$

in which an electron is transferred from a sodium atom to a chlorine atom with the eventual formation of a sodium ion and chloride ion. In sodium chloride the cohesive force is the electrostatic attraction between the oppositely charged ions.

The other case is where two electrons are shared by two nuclei so that the attractive force is predominant. The simplest example of such a *covalent bond* is in the diatomic molecule of hydrogen, H_2. In a somewhat

Combination of Atoms: The Formation of Diatomic Molecules 111

naïve way the nuclear (positive) and electronic (negative) charges may be written in the form:

$$+\genfrac{}{}{0pt}{}{-}{-}+$$

where the two electrons bind together the two nuclei.

More sophisticated versions of these bonding types will be developed in this and the following chapters.

7.3 The Formation of Molecular Orbitals in H_2^+ and H_2

In building up the electronic configurations of atoms the electrons were fed into atomic orbitals; the latter being defined by the three quantum numbers, n, l and m. When we come to the problems of describing molecular systems we can construct molecular orbitals from the atomic orbitals of the atoms present and fill them in accordance with the principles used in atom building (Section 6.3).

Consider the molecule of hydrogen, H_2. This contains two unipositive nuclei (protons) and two electrons. The equilibrium internuclear distance is observed to be 0·074 nm, this being the configuration of greatest stability resulting from the balancing of the attractive and repulsive forces in the system. From a consideration of the two 1s eigenfunctions (Figure 5.2) it may be concluded that an electron in the 1s orbital of one of the atoms would be considerably affected by the presence of the other nucleus.

The basis of molecular orbital theory is that in molecular systems one cannot consider electrons to be under the sole influence of their original nucleus. Rather, they are affected by *all* the nuclei present in the

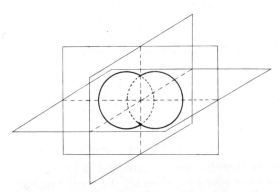

Figure 7.1. A representation of the overlapping of two 1s atomic orbitals

molecular system. In other words the effect of the several nuclei in a molecular system is to transform the atomic orbitals to molecular ones.

In the hydrogen molecule the two 1s atomic orbitals form *two* molecular orbitals. With the nuclei of the two hydrogen atoms 0·074 nm apart there is an appreciable amount of *overlap* of the two 1s atomic orbitals. A pictorial representation of this situation is shown in Figure 7.1.

To enquire into the ways in which molecular orbitals may be formed from these atomic orbitals it is necessary to look at the plots of the radial wave functions for the two 1s atomic orbitals. In Figure 7.2 the radial functions are plotted against the distance from their respective nuclei, these latter being again 0·074 nm apart. Again the overlapping of the wave functions is apparent.

Before proceeding to deal with the molecular orbitals of hydrogen, it is essential to consider a principle of fundamental importance. This is that when combining n atomic orbitals to form molecular orbitals, n of the latter must result. The reason for this is that n atomic orbitals can contain a maximum number of $2n$ electrons. When the molecular orbitals are formed this maximum occupancy must again be provided for. Since no orbital (atomic or molecular) may contain more than *two* electrons (Pauli Principle, Sections 6.2 and 3) it follows that:

> n atomic orbitals form n molecular orbitals

In the case of the hydrogen molecule the combination of the two 1s atomic orbitals must therefore result in the production of two molecular orbitals. The particular method used to produce the molecular orbitals from the atomic ones is known as the Linear Combination of Atomic Orbitals (LCAO) method.

If we represent the two 1s atomic orbitals of the hydrogen atoms by ψ_A and ψ_B we can produce two simple linear combinations of them. These are:

$$\psi_S = \psi_A + \psi_B \tag{7.2}$$

and

$$\psi_{AS} = \psi_A - \psi_B \tag{7.3}$$

The symmetric combination, ψ_S, represents a total ψ-value which is the sum of the two individual ψ-values for the atomic orbitals, ψ_A and ψ_B. This sum is shown in Figure 7.2, and it is of importance to note the effective summation occurring in the internuclear region.

Combination of Atoms: The Formation of Diatomic Molecules 113

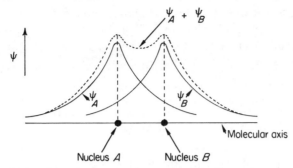

Figure 7.2. The symmetric overlapping of two $1s$ wave functions showing the separate wave functions together with their sum

It is the square of ψ which gives the probability density of finding electrons in any particular space element, and the square of ψ_S is shown in Figure 7.3, together with the quantity, $\psi_A{}^2 + \psi_B{}^2$.

It is to be noted that $\psi_S{}^2$ is greater than $\psi_A{}^2 + \psi_B{}^2$ in the internuclear region. This means that, compared to the case where the atomic orbitals do not interact, there is a greater probability of finding electrons in the internuclear region. This is just the condition for the effective binding together of the positive nuclei. Whether one or two electrons are present in ψ_S the situations correspond to overall attractive or *bonding* states:

$$+ \quad - \quad + \quad \text{(one electron)}$$

or

$$+ \quad \begin{matrix}-\\-\end{matrix} \quad + \quad \text{(two electrons)}$$

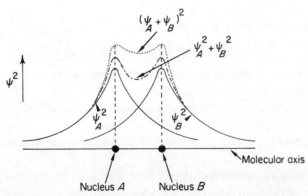

Figure 7.3. Plots of $\psi_A{}^2$, $\psi_B{}^2$, $\psi_A{}^2 + \psi_B{}^2$ and $(\psi_A + \psi_B)^2$ for the hydrogen molecule

Mathematical justification for the above statements comes from the consideration of the probability of finding electrons in the ψ_S orbital. So far the normalization factors (Section 4.6) have been omitted from the wave functions. We may now write equation (7.2) as

$$\psi_S = N_b(\psi_A + \psi_B) \tag{7.4}$$

where N_b is the normalizing factor which ensures that when ψ_S^2 is integrated over all space the result is unity, viz.:

$$\int_0^\infty \psi_S^2 \, d\tau = 1 \cdot 0 \tag{7.5}$$

so that:

$$N_b^2 \int_0^\infty (\psi_A + \psi_B)^2 \, d\tau = 1 \cdot 0 \tag{7.6}$$

Expanding the integral (7.6), we get:

$$N_b^2 \int_0^\infty \psi_A^2 \, d\tau + N_b^2 \int_0^\infty \psi_B^2 \, d\tau + 2N_b^2 \int_0^\infty \psi_A \cdot \psi_b \, d\tau = 1 \cdot 0 \tag{7.7}$$

We may consider that the functions, ψ_A and ψ_B, are separately normalized so that

$$\int_0^\infty \psi_A^2 \, d\tau = 1 \cdot 0 \tag{7.8}$$

and

$$\int_0^\infty \psi_B^2 \, d\tau = 1 \cdot 0 \tag{7.9}$$

by which it is possible to simplify the integral (7.7) to

$$N_b^2(1 + 1 + 2S) = 1 \cdot 0 \tag{7.10}$$

where S represents what is known as the *overlap integral*:

$$\int_0^\infty \psi_A \cdot \psi_B \, d\tau$$

and represents the extent of overlap of the two original atomic orbitals (Figure 7.1).

By rearranging equation (7.10) we find that:

$$N_b = \frac{1}{[2(1+S)]^{1/2}} \tag{7.11}$$

Combination of Atoms: The Formation of Diatomic Molecules

so that the normalized molecular wave function is

$$\psi_S = \frac{1}{[2(1+S)]^{1/2}}(\psi_A + \psi_B) \tag{7.12}$$

where ψ_A and ψ_B are the separately normalized atomic wave functions.

We can now apply the same procedures to the *antisymmetric* combination of the two atomic wave functions. The two radial functions are arranged antisymmetrically in Figure 7.4 and their sum (i.e. $(\psi_A) + (-\psi_B)$) is shown in Figure 7.5. The total ψ-value at a distance half-way between the two nuclei is zero (cf. Figure 7.2).

The antisymmetric wave function may be normalized by writing:

$$\psi_{AS} = F_{ab}(\psi_A - \psi_B) \tag{7.13}$$

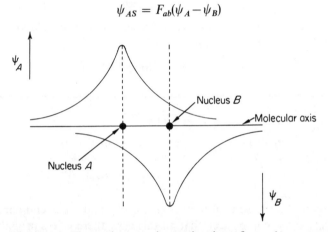

Figure 7.4. The antisymmetric overlapping of two $1s$ wave functions

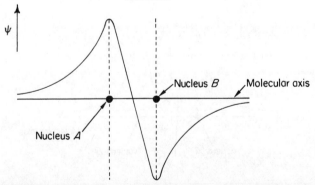

Figure 7.5. The resultant of the antisymmetric combination of the $1s$ wave functions

where F_{ab} is the normalizing factor, and by arranging the integral of ψ_{AS}^2 over all space to be unity, viz.:

$$\int_0^\infty \psi_{AS}^2 \cdot d\tau = 1\cdot 0 \qquad (7.14)$$

or

$$F_{ab}^2 \int_0^\infty (\psi_A - \psi_B)^2 \cdot d\tau = 1\cdot 0 \qquad (7.15)$$

which may be expanded to read:

$$F_{ab}^2 \left[\int_0^\infty \psi_A^2 \cdot d\tau + \int_0^\infty \psi_B^2 \cdot d\tau - 2 \int_0^\infty \psi_A \cdot \psi_B \cdot d\tau \right] = 1\cdot 0 \qquad (7.16)$$

If we again consider ψ_A and ψ_B to be separately normalized (integrals 7.8 and 7.9) we can rewrite equation (7.16) as

$$F_{ab}^2 (1 + 1 - 2S) = 1\cdot 0 \qquad (7.17)$$

or

$$F_{ab} = \frac{1}{[2(1-S)]^{1/2}} \qquad (7.18)$$

so that the normalized antisymmetric wave function is

$$\psi_{AS} = \frac{1}{[2(1-S)]^{1/2}} (\psi_A - \psi_B) \qquad (7.19)$$

In Figure 7.6 the sum of the squares of ψ_A and ψ_B is compared with $(\psi_A - \psi_B)^2$ and it can be seen that, in this case, there is less probability of

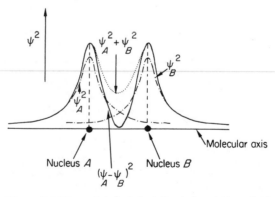

Figure 7.6. Plots of ψ_A^2, ψ_B^2, $\psi_A^2 + \psi_B^2$ and $(\psi_A - \psi_B)^2$ for the hydrogen molecule

Combination of Atoms: The Formation of Diatomic Molecules

finding electrons in the internuclear region than in the case of the two non-interacting atomic orbitals. In a simple way it could be said that if two electrons occupied the ψ_{AS} orbital the most probable distribution would be

$$- \quad + \quad + \quad -$$

which is a repulsive or *antibonding* situation.

We can now continue with a consideration of the relative energies of the orbitals, ψ_A, ψ_B, ψ_S and ψ_{AS}.

Starting with the wave equation (Section 4.2) in the form:

$$H\psi = E\psi \tag{7.20}$$

we can multiply each side by ψ and obtain:

$$\psi H\psi = \psi E\psi \tag{7.21}$$

and we can then integrate both sides over all space:

$$\int_0^\infty \psi H\psi \, . \, d\tau = \int_0^\infty \psi E\psi \, . \, d\tau \tag{7.22}$$

this being a mathematical device to allow the calculation of E. Since E is a constant (unlike H which is an operator) it may be taken from the integral on the right-hand side of equation (7.22) which then may be written:

$$\int_0^\infty \psi H\psi \, . \, d\tau = E \int_0^\infty \psi^2 \, . \, d\tau \tag{7.23}$$

which gives

$$E = \int_0^\infty \psi H\psi \, . \, d\tau \tag{7.24}$$

since, if ψ is normalized (Section 4.6) the value of the integral on the right-hand side of equation (7.23) is unity. Equation (7.24) may be applied to the symmetric orbital, ψ_S, whose energy is given by:

$$E_b = N_b^2 \int_0^\infty (\psi_A + \psi_B) H(\psi_A + \psi_B) \, . \, d\tau$$

$$= N_b^2 \int_0^\infty \psi_A H\psi_A \, . \, d\tau + N_b^2 \int_0^\infty \psi_B H\psi_B \, . \, d\tau$$

$$+ 2N_b^2 \int_0^\infty \psi_A H\psi_B \, . \, d\tau \tag{7.25}$$

The first and second integrals in equation (7.25) represent the energies of electrons in the 1s atomic orbital of an isolated hydrogen atom and are identical. These two quantities may be represented by α where

$$\alpha = N_b^2 \int_0^\infty \psi_A H \psi_A \, d\tau = N_b^2 \int_0^\infty \psi_B H \psi_B \, d\tau \tag{7.26}$$

The quantity α forms the reference with which the energies of the two molecular orbitals, ψ_S and ψ_{AS}, may be compared. The integrals, α, are known as coulomb integrals and represent the energy which is *released* when an electron initially separated from the nucleus by an infinite distance occupies the 1s atomic orbital of the hydrogen atom. The energy represented by α is therefore, by convention, a *negative* quantity compared to the energy zero where the electron and nucleus are infinitely separated. The energy required to completely remove the electron from the 1s orbital of a hydrogen atom is, by definition, the ionization potential (Section 2.5) and is equal to $-\alpha$.

The third integral in equation (7.25) may be put equal to a quantity of energy β, which is known as the resonance integral, and is also a negative quantity. It represents the extra energy released due to the formation of the molecular orbital.

Equation (7.25) may be rewritten in terms of the quantities α and β as:

$$E_b = N_b^2 \alpha + N_b^2 \alpha + 2N_b^2 \beta = 2N_b^2(\alpha + \beta) \tag{7.27}$$

and by putting in the value of N_b as given by equation (7.11) this becomes:

$$E_b = \frac{\alpha + \beta}{1 + S} \tag{7.28}$$

Equation (7.28) may be rearranged in the following manner:

$$E_b = \frac{\alpha}{1+S} + \frac{\beta}{1+S} \tag{7.29}$$

and by adding αS to the first fraction and by subtracting αS from the second this becomes:

$$E_b = \frac{\alpha + \alpha S}{1+S} + \frac{\beta - \alpha S}{1+S} \tag{7.30}$$

which simplifies to:

$$E_b = \alpha + \frac{\beta - \alpha S}{1+S} \tag{7.31}$$

Combination of Atoms: The Formation of Diatomic Molecules

Carrying out a similar treatment for the antisymmetric molecular orbital, ψ_{AS}, we have for the electron energy:

$$E_{ab} = F_{ab}^2 \int_0^\infty (\psi_A - \psi_B) H (\psi_A - \psi_B) \, d\tau$$

$$= F_{ab}^2 \int_0^\infty \psi_A H \psi_A \, d\tau + F_{ab}^2 \int_0^\infty \psi_B H \psi_B \, d\tau$$

$$- 2 F_{ab}^2 \int_0^\infty \psi_A H \psi_B \, d\tau \tag{7.32}$$

This may be written as:

$$E_{ab} = F_{ab}^2 \alpha + F_{ab}^2 \alpha - 2 F_{ab}^2 \beta$$

$$= 2 F_{ab}^2 (\alpha - \beta) \tag{7.33}$$

Putting in the value of F_{ab} given by equation (7.18) the expression becomes

$$E_{ab} = \frac{\alpha - \beta}{1 - S} \tag{7.34}$$

which may be expanded:

$$E_{ab} = \frac{\alpha}{1 - S} - \frac{\beta}{1 - S} \tag{7.35}$$

The factor αS is subtracted from both fractions in equation (7.35) giving

$$E_{ab} = \frac{\alpha - \alpha S}{1 - S} - \frac{\beta - \alpha S}{1 - S} \tag{7.36}$$

which simplifies to:

$$E_{ab} = \alpha - \frac{\beta - \alpha S}{1 - S} \tag{7.37}$$

We are now in a position to draw a diagram of the relative energies of the orbitals, ψ_A, ψ_B, ψ_S and ψ_{AS}. The atomic orbitals, ψ_A and ψ_B, have energies equal to α (equation 7.26) and this energy can be made the reference point for the energies, E_b and E_{ab}. The symmetric orbital combination, ψ_S, produces an electron energy, E_b, which is greater than α by the term, $(\beta - \alpha S)/(1 + S)$, (equation 7.31), both terms being negative quantities. This means that compared to an isolated hydrogen atom, it is more difficult to remove an electron from this orbital. The orbital energy is therefore lower than that in the isolated hydrogen atom. For this reason the orbital, ψ_S, is known as a *bonding* molecular orbital. An electron is more stable in a bonding orbital than it is in the separated atomic state.

The antisymmetric orbital combination, ψ_{AS}, has an energy which is less than α by the term, $(\beta - \alpha S)/(1 - S)$, (equation 7.37), and an electron in this orbital will be less stable than in the isolated atomic state. This orbital is, in consequence, known as an *antibonding* molecular orbital.

The situation is summarized in Figure 7.7 where the energies of the bonding and antibonding molecular orbitals are compared with the

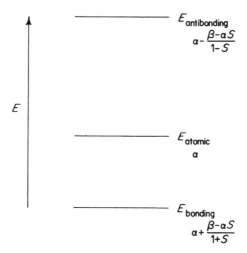

Figure 7.7. The relative energies of the eigenfunctions, ψ_S (bonding orbital) and ψ_{AS} (antibonding orbital), as compared with the energy of the atomic orbitals, ψ_A and ψ_B

energy of the atomic orbitals from which the molecular orbitals were constructed.

For bonding to occur the atomic orbitals must overlap, i.e. S must be non-zero. A common approximation, however, is to assume S to be zero in which case E_b becomes equal to $\alpha + \beta$ and E_{ab} becomes equal to $\alpha - \beta$. If, *in fact*, S is zero then, since the orbitals concerned do not overlap, the integrals,

$$2N_b^2 \int_0^\infty \psi_A H \psi_B \, d\tau \quad \text{and} \quad 2F_{ab}^2 \int_0^\infty \psi_A H \psi_B \, d\tau$$

in equations (7.25) and (7.32) respectively are also zero. This means that

Combination of Atoms: The Formation of Diatomic Molecules

β is zero and $E_b = E_{ab} = \alpha$. In other words, bonding will not take place since there is no energetic advantage in such an occurrence.

As two hydrogen atoms are brought together there is a distance of separation where S becomes non-zero and, as the internuclear distance decreases, S increases. This means that E_b, $E_{\text{atomic}}(\alpha)$ and E_{ab}, which are all degenerate at $r = \infty$, separate, and the gap between E_{ab} and E_b increases with increasing S.

The variation of energy with internuclear distance is represented in Figure 7.8 which also takes into account the internuclear repulsion which

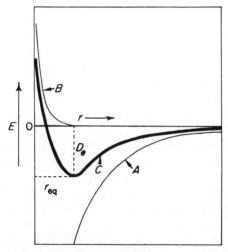

Figure 7.8. Plots of potential energy versus internuclear distance for a diatomic molecule

operates at small internuclear distances. Curve A represents the lowering of energy as two hydrogen atoms are brought together from infinity (the latter position is taken to be the zero of energy), curve B represents the short range effect of internuclear repulsion where the energy increases rapidly as the internuclear distance decreases, and curve C represents the *sum* of curves A and B. The minimum in curve C occurs at an internuclear distance known as the equilibrium separation, r_{eq}. With the latter internuclear separation the energy of the molecule is D_e below the zero (infinite separation). D_e is known as the electronic dissociation energy of the molecule. It does not represent the actual bond dissociation energy, D, since all molecules possess *zero point energy* due to vibrational motion. The zero point energy of a diatomic molecule is given by $\frac{1}{2}h\omega_0$ (this is derived by putting $v = 0$ in equation 2.19) where ω_0 is the fundamental vibration frequency. The bond dissociation energy is equal to $D_e - \frac{1}{2}h\omega_0$ and represents the energy which has to be *used* in order to break the bond between the two atoms completely. The quantity, D, is also the amount of energy *released* when the molecule is *formed* from the two separated atoms.

The coulombic attractive force between the electrons and the nuclei in the hydrogen molecule which produces the lowering of energy, E_b, as the two nuclei come closer together is balanced eventually by the coulombic internuclear repulsion. This balancing occurs when the nuclei are 0·074 nm apart.

The bonding and antibonding molecular orbitals, together with the component 1s atomic orbitals, are shown pictorially in Figure 7.9. The consequence of the bonding theory outlined in this section is that if one electron is placed in the ψ_S orbital (bonding) it will bond together the two hydrogen nuclei in forming H_2^+, the hydrogen molecule-ion. H_2^+ would be more stable than $H + H^+$ by the amount of energy equal to the difference, $E_{atomic} - E_b$. In an electrical discharge through low pressure hydrogen gas the spectrum of H_2^+ has been observed, and from the spectrum it has been estimated that the internuclear distance is 0·106 nm and that the bond dissociation energy (the energy required to cause the dissociation, $H_2^+ \rightarrow H + H^+$) is 64 kcal.mole^{-1} (268 kJ.mole^{-1}).

Again, if two electrons are allowed to occupy the bonding orbital (and they can do so if they possess opposed spins—see Section 6.2) they will

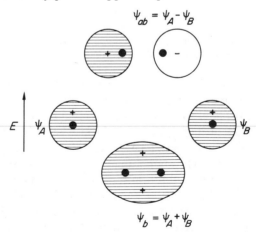

Figure 7.9. Pictorial representations of the two 1s atomic orbitals and the products of their symmetric and antisymmetric combination to give bonding and antibonding orbitals

Combination of Atoms: The Formation of Diatomic Molecules

bind the two nuclei together in forming the hydrogen molecule, H_2. The effect of the second electron is to increase the binding compared to that in H_2^+ and the internuclear distance decreases to 0·074 nm while the bond dissociation energy increases to 104 kcal.mole^{-1} (435 kJ.mole^{-1}). The fact that the two electrons do not produce a bond dissociation energy which is twice that in H_2^+ is an indication that there is a considerable amount of electron–electron repulsion. This is not, however, great enough to prevent a considerably more stable molecule from being produced.

The energies of the atomic and molecular orbitals of Figure 7.7 are those for single electron occupancy. This means that it should not be expected that two electrons in the bonding molecular orbital would have *twice* the bonding effect that a single electron would have. The reason for this is that diagrams such as Figure 7.7 take no account of electron–electron repulsion. The contribution to the total energy of the system from internuclear repulsion is also neglected. The effects of internuclear repulsion in H_2^+ and H_2, and of interelectronic repulsion in H_2, are important quantities, and it is, to a large extent, fortuitous that the bond energy of H_2 is roughly twice that of H_2^+. Linnett has shown that the electronic binding energy in H_2 is only 10 per cent greater than that in H_2^+. In his own words, '... the particular significance of the electron pair bond appears to reside in the feature that two electrons are a little better than one ...'.

The two molecular orbitals produced by combining two $1s$ atomic orbitals are not spherically symmetrical but are cylindrically symmetrical with respect to the molecular axis. They are known as sigma (σ) orbitals and the bond formed by the two-electron occupation of the bonding orbital, ψ_S, is known as a sigma (σ) bond. The bonding orbital may be symbolized as a σ_{1s} orbital indicating that it has been formed from $1s$ atomic orbitals. The antibonding orbital, ψ_{AS}, may be written as σ_{1s}^*, the asterisk indicating its antibonding character.

As was pointed out earlier in this section, provision must be made in the molecular orbitals of a system for the maximum number of electrons capable of being accommodated in the atomic orbitals from which the molecular ones are produced. In the case already discussed for the combinations of two $1s$ atomic orbitals, it is possible to apply this theory to species involving three and four electrons.

The helium molecule ion, He_2^+, possesses three electrons. Two of these would normally occupy the σ_{1s} bonding orbital while the third electron would have to occupy the σ_{1s}^* antibonding level. The electronic configuration of He_2^+ would be $(\sigma_{1s})^2(\sigma_{1s}^*)^1$. It would be expected that the presence of *one* antibonding electron would offset to more than 50 per cent (see

Figure 7.7) the bonding effects of the two electrons in the bonding orbital. The electron in the antibonding orbital weakens the bonding effect because of its antibonding nature and because it contributes to greater interelectronic repulsion. The helium molecule-ion does exist in a helium discharge and its bond dissociation energy is found to be 71 kcal.mole^{-1} (297 kJ.mole^{-1}). The internuclear separation is 0·108 nm. These are of the same order as those for H_2^+ and are qualitatively consistent with the theory. In both H_2^+ and He_2^+ the excess of bonding over antibonding electrons is one.

The electronic configuration of the helium diatomic molecule would be $(\sigma_{1s})^2(\sigma_{1s}^*)^2$, in this case there being equal numbers of bonding and antibonding electrons. The theory would therefore predict that the bond dissociation energy of He_2 would be zero and, in fact, shows that such a molecule cannot be any more stable than a pair of isolated helium atoms. The molecule has not been observed with this particular electronic configuration.

7.4 Application of Molecular Orbital Theory to the Homonuclear Diatomic Molecules of the First Short Period Elements

Similar considerations to those dealt with in the previous section apply to the overlapping of 2s atomic orbitals. In such a case the molecular orbitals produced may be written as σ_{2s} and σ_{2s}^*. The first element in the first short period of the periodic classification (Figure 6.10) is lithium. The lithium atom has the electronic configuration $1s^2\,2s^1$. The vapour of the element does contain some Li_2 molecules. The internuclear separation is 0·267 nm and the bond dissociation energy is 25 kcal.mole^{-1} (104·5 kJ.mole^{-1}). At such a distance apart the 1s atomic orbitals of the lithium

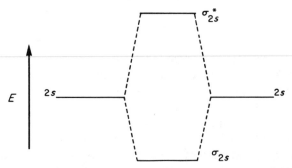

Figure 7.10. The energies of the σ_{2s} and σ_{2s}^* molecular orbitals compared to the energy of the 2s atomic orbitals from which the former were constructed

atoms would not overlap to any significant extent (the $1s$ orbital in lithium is considerably smaller than that in hydrogen because of the increased nuclear charge—see Table 5.1), and, in consequence, the $1s^2$ pairs of electrons on the lithium atoms may be considered to be localized and non-bonding (lone pairs). The more diffuse $2s$ orbitals form a molecular orbital system and this is shown in Figure 7.10. The two $2s$ electrons in the separated 2Li state would occupy the σ_{2s} bonding orbital in the Li_2 molecule whose ground state may be written as:

$$Li_2\ KK(\sigma_{2s})^2$$

the KK being symbolic of the two $1s^2$ lone pairs. (The $1s$ orbital is sometimes referred to as the K shell of an atom.)

Beryllium with the configuration, $1s^2\ 2s^2$, would not be expected to form a diatomic molecule since one would have the configuration,

$$Be_2\ KK(\sigma_{2s})^2(\sigma_{2s}^*)^2$$

and there would be no resultant bonding. No such molecule exists.

Boron has a ground state configuration, $1s^2\ 2s^2\ 2p^1$, and the diatomic molecule, B_2, is observed to have an equilibrium internuclear separation of 0·159 nm and a bond dissociation energy of 69 kcal.mole^{-1} (289 kJ.mole^{-1}). It is necessary to deal with the overlap situations of two $2p$ orbitals. We can consider these to be directed along the molecular axis with their coordinate systems being chosen so that the orbitals are arranged for overlap as shown in Figure 7.11. The bonding (symmetric) combination would be given by

$$\psi_p = \psi_{2p}(A) + \psi_{2p}(B) \tag{7.38}$$

and the antibonding (antisymmetric) one by

$$\psi_p{}^* = \psi_{2p}(A) - \psi_{2p}(B) \tag{7.39}$$

The molecular orbitals may be described as σ_{2p} and σ_{2p}^* respectively and, if necessary, the molecular axis may be specified. For example, if the molecular axis is arbitrarily taken to be the z axis, then the two orbitals would be written as σ_{2p_z} and $\sigma_{2p_z}^*$.

The corresponding pictorial representations are shown in Figure 7.11 together with an indication of the relative energies of the orbitals concerned.

There is also the possibility of combining the other $2p$ orbitals to form molecular orbitals but these cannot overlap to form sigma orbitals (as can the $2p_z$ orbitals). The $2p_x$ and $2p_y$ orbitals are directed perpendicularly to each other and to the molecular axis (taken arbitrarily as the z axis).

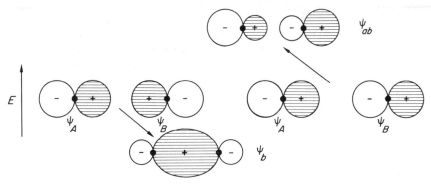

Figure 7.11. Pictorial representations of the formation of the bonding and antibonding molecular orbitals by the symmetric and antisymmetric combinations of two p_z orbitals. The relative energies of the orbitals are indicated

It is possible for the $2p_x$ and $2p_y$ sets of orbitals to overlap 'sideways' as is shown in Figure 7.12. The quantum mechanical descriptions of the molecular orbitals may, for our purposes, be taken to be:

$$\psi_{\pi_x} = \psi_{2p_x}(A) + \psi_{2p_x}(B) \tag{7.40}$$

$$\psi_{\pi_x}^* = \psi_{2p_x}(A) - \psi_{2p_x}(B) \tag{7.41}$$

$$\psi_{\pi_y} = \psi_{2p_y}(A) + \psi_{2p_y}(B) \tag{7.42}$$

$$\psi_{\pi_y}^* = \psi_{2p_y}(A) - \psi_{2p_y}(B) \tag{7.43}$$

The bonding and antibonding combinations of p orbitals described by equations (7.40), (7.41), (7.42) and (7.43) and as shown in Figure 7.12 are not cylindrically symmetrical about the molecular axis. They are not sigma orbitals. Viewed down the molecular axis they have the appearance of p atomic orbitals. That is, they possess a nodal plane (i.e. a plane where ψ and ψ^2 are zero) which contains the molecular axis. The nodal plane for the $2p_x$ combination is the yz plane and that for the $2p_y$ combination is the xz plane. Such orbitals with one nodal plane (containing the molecular axis) are known as pi (π) orbitals.

Since the p_x and p_y orbitals are degenerate and undertake the same type of molecular orbital formation, then the π_x and π_y orbitals are degenerate, as are the π_x^* and π_y^* pair.

The energies of the degenerate pair of π bonding orbitals and that of the σ_{2p} bonding orbital vary with the nuclear charge of the atoms involved in the diatomic molecule. This factor alters the order of filling of the orbitals. Further details of this subject are dealt with in Appendix III.

In the molecule, B_2, there are two $1s^2$ lone pairs and the remaining six

Combination of Atoms: The Formation of Diatomic Molecules

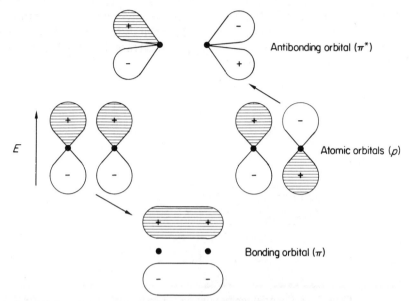

Figure 7.12. Pictorial representations of the formation of the bonding and antibonding molecular orbitals by the symmetric and antisymmetric combinations of two p_x (or p_y) atomic orbitals. The relative energies of the orbitals are indicated

electrons occupy the lowest three molecular levels. These are the σ_{2s}, σ_{2s}^* and the doubly degenerate π_{2p} levels. The π_{2p} levels in this case are lower in energy than the σ_{2p} orbital. With the σ_{2s} and σ_{2s}^* orbitals being occupied by pairs of electrons, this leaves two electrons to be accommodated by the π_{2p} orbitals. In accordance with Hund's Rules (Section 6.3) the two orbitals (π_{2p_x} and π_{2p_y}) are singly occupied and the resulting electronic state is a triplet. The electronic configuration of the B_2 molecule may be written as:

$$B_2 \; KK(\sigma_{2s})^2 \, (\sigma_{2s}^*)^2 \, (\pi_{2p_x})^1 \, (\pi_{2p_y})^1$$

In the C_2 molecule the degenerate π_{2p} levels are still lower than the σ_{2p} bonding level and its electronic configuration is

$$C_2 \; KK(\sigma_{2s})^2 \, (\sigma_{2s}^*)^2 \, (\pi_{2p_x})^2 \, (\pi_{2p_y})^2$$

The C_2 molecule has an equilibrium internuclear separation of 0·124 nm and a bond dissociation energy of 113 kcal.mole^{-1} (473 kJ.mole^{-1}). The bond is stronger than that in the B_2 molecule. The data for C_2, compared with those for B_2, are consistent with there being an extra pair of bonding electrons in the molecule. When electrons occupy pi bonding orbitals the bond thus produced is described as a pi bond. As has been mentioned above, there are two sets of pi molecular orbitals (chosen to be in the x and y directions), and the two bonding pi orbitals are shown in Figure 7.13.

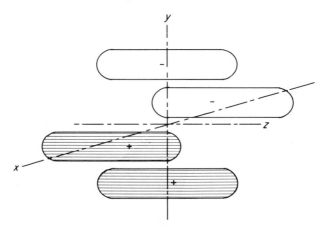

Figure 7.13. Pictorial representation of two mutually perpendicular pi molecular orbitals (those formed by the symmetric overlap of pairs of p_x and p_y orbitals respectively)

In the nitrogen molecule, N_2, there are two extra electrons to be accommodated as compared to C_2. These two electrons occupy the lowest available molecular orbital which is the σ_{2p_z} bonding orbital. The electronic configuration is thus:

$$N_2 \; KK(\sigma_{2s})^2 \, (\sigma_{2s}^*)^2 \, (\pi_{2p_x})^2 (\pi_{2p_y})^2 (\sigma_{2p_z})^2$$

There are six bonding electrons in excess of the antibonding ones and compared to C_2 there should be a decrease in the internuclear separation and an increase in the bond dissociation energy. The observed values are 0·109 nm and 225 kcal.mole^{-1} (940 kJ.mole^{-1}) respectively for the two quantities, these being consistent with expectation.

When the nitrogen molecule is singly ionized (as may occur in an electrical discharge through the gas, or in a mass spectrometer) the electron which is most easily removed is the one in the highest energy orbital (σ_{2p_z}) and since this is a bonding orbital it would be expected that in N_2^+ there would be an increased internuclear separation and a decreased bond dissociation energy compared to the neutral molecule, N_2. The data are consistent with such predictions, the internuclear separation of N_2^+ being 0·112 nm and its bond dissociation energy being 201 kcal.mole^{-1} (840 kJ.mole^{-1}).

Combination of Atoms: The Formation of Diatomic Molecules

In the oxygen molecule, O_2, there are sixteen electrons. Four of these form the two $1s^2$ lone pairs and another four will fill the σ_{2s} and σ_{2s}^* orbitals, leaving eight electrons to be accommodated in orbitals of higher energy. The different nuclear charge produces a different order of filling of the molecular orbitals (see Appendix III) which in the oxygen molecule is

$$\sigma_{2s} < \sigma_{2s}^* < \sigma_{2p_z} < \pi_{2p_x} = \pi_{2p_y} < \pi_{2p_x}^* = \pi_{2p_y}^*$$

The electronic configuration of the oxygen molecule is therefore

$$O_2 \; KK(\sigma_{2s})^2 (\sigma_{2s}^*)^2 (\sigma_{2p_z})^2 (\pi_{2p_x})^2 (\pi_{2p_y})^2 (\pi_{2p_x}^*)^1 (\pi_{2p_y}^*)^1$$

in which the doubly degenerate $\pi_{2p_x}^*$ and $\pi_{2p_y}^*$ orbitals are singly occupied in accordance with Hund's Rules (Section 6.3). The electronic state of the molecule is a triplet (cf. B_2) in which the two unpaired electrons have parallel spins.

The predictions of molecular orbital theory for O_2 are that it should be paramagnetic (since there are unpaired electrons present), and compared to nitrogen there should be a longer bond and a lower bond dissociation energy. The latter two predictions are due to there being an excess of only four bonding electrons over the non-bonding ones, as compared to six in the case of the nitrogen molecule.

The observed facts are (for O_2) a bond length of 0·121 nm and a bond dissociation energy of 118 kcal.mole^{-1} (493 kJ.mole^{-1}), and the molecule is paramagnetic. The facts are qualitatively consistent with the theoretical predictions.

Furthermore, data concerning the O_2^+ ion support the idea that the highest energy electrons in O_2 are of an antibonding nature. If an antibonding electron is removed from O_2 in the formation of O_2^+ the latter should have a shorter bond with a higher dissociation energy than does O_2.

The oxygen molecule-ion, O_2^+, does exist in the compound, $O_2^+ PtF_6^-$, as well as in the gaseous phase in an electrical discharge through oxygen gas. Its bond length in the gaseous state is found to be 0·112 nm and its bond dissociation energy if 149 kcal.mole^{-1} (623 kJ.mole^{-1}). Moreover the compound, $O_2^+ PtF_6^-$, is paramagnetic, as would be expected for O_2^+ (one unpaired electron) and PtF_6^- (also one unpaired electron). The facts once again fully justify the theory.

In the molecule of fluorine, F_2, the antibonding pi orbitals are fully occupied and the electronic configuration is

$$F_2 \; KK(\sigma_{2s})^2 (\sigma_{2s}^*)^2 (\sigma_{2p_z})^2 (\pi_{2p_x})^2 (\pi_{2p_y})^2 (\pi_{2p_x}^*)^2 (\pi_{2p_y}^*)^2$$

There is an excess of two bonding electrons over the antibonding ones and as would be expected the bond in the molecule is both longer and weaker than that in O_2. The bond length is observed to be 0·144 nm and the bond dissociation energy is 37 kcal.mole^{-1} (155 kJ.mole^{-1}).

The molecule, Ne_2, does not exist, which is consistent with the prediction that it would contain an equal number of bonding and antibonding electrons.

A summary of the predictions of molecular orbital theory and the corresponding physical data for the molecules discussed in this section is shown in Table 7.1.

Table 7.1. Predictions of Molecular Orbital Theory and Their Correlation with Experimental Observation for Some Diatomic Molecules

	Molecular orbital predictions				Experimental observations		
						Bond dissociation energy	
Molecule	Number of bonding electrons	Number of anti-bonding electrons	Number of excess bonding electrons	Bond order	Bond length (nm)	(kcal.-mole^{-1})	(kJ.-mole^{-1})
Li_2	2	0	2	1	0·267	25	105
Be_2	2	2	0	0	—	—	—
B_2	4	2	2	1	0·159	69	289
C_2	6	2	4	2	0·124	113	473
N_2	8	2	6	3	0·109	225	940
N_2^+	7	2	5	$2\frac{1}{2}$	0·112	201	840
O_2	8	4	4	2	0·121	118	493
O_2^+	8	3	5	$2\frac{1}{2}$	0·112	149	623
F_2	8	6	2	1	0·144	37	155
Ne_2	8	8	0	0	—	—	—

If, as in the molecule of hydrogen, it is considered that two bonding electrons form a covalent bond, it is possible to assign what are known as *bond orders* to the bonds in the molecules in Table 7.1. The order of a bond comprising *two* bonding electrons is *one* and, for diatomic molecules, we can define bond order as being one half of the number of excess bonding electrons in the bond in question. The bond orders corresponding to the bonds in the molecules discussed are shown in Table 7.1.

Combination of Atoms: The Formation of Diatomic Molecules

The correlations between bond orders and bond lengths and dissociation energies are not perfect, but this is not surprising since the atoms concerned possess different values of nuclear charge. The extent of the correlation is extremely encouraging when it is considered that the applications of wave-mechanical principles to systems with more than one electron are approximate.

7.5 The Hybridization of Atomic Orbitals

In Section 7.3 it was shown that the stabilization of bonding molecular orbitals with respect to the free atoms was increased by increasing the value of the overlap integral, S, where

$$S = \int_0^\infty \psi_A \cdot \psi_B \cdot d\tau \tag{7.44}$$

In Section 7.4 we have dealt with the overlap of pure s orbitals to give molecular orbitals and of pure p orbitals to give sigma and pi molecular orbitals.

It can be shown that greater effective bonding may be achieved by using *hybrid atomic orbitals* to construct the molecular system. A hybrid orbital is one which is neither pure s nor pure p (nor d and f) but is some mixture of several pure atomic orbitals. This process of mixing atomic orbitals is known as *hybridization* and can be exemplified by considering the mixing together of an s orbital with a p orbital.

Wave-mechanically this is done by making linear combinations of the atomic wave functions, viz.:

$$\psi_{sp}(A) = \psi_s + \psi_p \tag{7.45}$$

$$\psi_{sp}(B) = \psi_s - \psi_p \tag{7.46}$$

The formation of the two hybrid sp orbitals, $\psi_{sp}(A)$ and $\psi_{sp}(B)$ is depicted in Figure 7.14.

The two sp hybrid orbitals are centred, of course, on the same atom and the two are superimposed in Figure 7.15.

The 'positive' lobes are larger than those of the pure s orbital. Thus when such hybrid orbitals are used in the formation of molecular orbitals, the overlap integral is larger than if a pure s orbital were used. Hybrid atomic orbitals, together with the orbitals of the other atoms which are bonding to the atom under consideration, form a better approximate

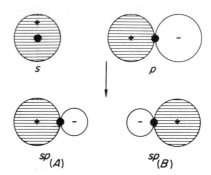

Figure 7.14. Pictorial representation of the formation of two sp hybrid atomic orbitals from pure s and p atomic orbitals

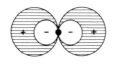

Figure 7.15. Pictorial representation of two sp hybrid orbitals

description (LCAO) of the resulting molecule than the use of pure atomic orbitals.

The case of hybridization just discussed is fairly straightforward but it should be pointed out that there are limitations to the extent to which hybridization of atomic orbitals can be incorporated into bonding schemes. For two or more atomic orbitals to take part in the formation of hybrid orbitals they should not differ in energy by any large amount. As far as the $2s$ and $2p$ orbitals are concerned, the energy gap between the two increases as the first short period is traversed from lithium to neon (Figure 5.12, see also Figure 8.18). In consequence, sp hybridization is more important with the early members of the period and becomes less and less important as neon is approached.

To allow for the unequal mixing of s and p orbitals, equations (7.45) and (7.46) must be modified to read:

$$\psi_{sp}(A) = \psi_s + \lambda\psi_p \qquad (7.47)$$

$$\psi_{sp}(B) = \lambda\psi_s - \psi_p \qquad (7.48)$$

where λ is a constant term.

The term λ is incorporated into equations (7.47) and (7.48) to allow for unequal contributions of the s and p atomic orbitals to the two hybrid ones. If they make equal contributions to both hybrid orbitals, then λ

Combination of Atoms: The Formation of Diatomic Molecules 133

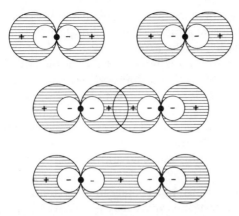

Figure 7.16. The formation of a bonding orbital by the symmetric combination of two *sp* hybrid atomic orbitals. Two other *sp* orbitals are also indicated which may accommodate lone pairs of electrons

has a value of unity and equations (7.47) and (7.48) reduce to equations (7.45) and (7.46) respectively.

The sigma bonds between the atoms in the diatomic molecules dealt with in Section 7.4 are therefore to be considered as being produced by the overlapping of two hybrid orbitals. The extent of the hybridization depends upon the 2*p*–2*s* energy gap, being small for large energy gaps.

The production of an *sp*–*sp* sigma bond from the overlap of two *sp* hybrid orbitals is shown in Figure 7.16, together with the two *sp* hybrid orbitals which do not interact and may accommodate lone pairs of electrons.

As far as diatomic molecules are concerned, the hybridization of atomic orbitals allows molecular orbitals to be formed which possess greater values of the overlap integral, S, than would be the case when the pure atomic orbitals are used.

The hybridization of atomic orbitals in polyatomic molecules is dealt with in Chapter 8.

7.6 The Hydrogen Fluoride Molecule

Hydrogen fluoride, HF, is a *heteronuclear* diatomic molecule, being composed of two *unlike* atoms. This being so, the theoretical treatment of the bonding between the two atoms requires some modification of that given to *homonuclear* diatomic molecules in Section 7.4.

The hydrogen atom in its ground state possesses the configuration $1s^1$ and that of the fluorine atom is (He) $2s^2\ 2p_x^2\ 2p_y^2\ 2p_z^1$, a single electron occupying, arbitrarily, the $2p_z$ orbital.

A sigma type of overlap is possible between the two singly occupied orbitals ($1s$ and $2p_z$) to give a bonding and antibonding pair of molecular orbitals. The two orbitals are of different energies which is apparent from the ionization potentials of the two atoms. To remove the electron from the $1s$ orbital of a hydrogen atom requires 13·6 ev while the first ionization potential of the fluorine atom is 17·42 ev. The latter value corresponds to the energy required to remove an electron from one of the doubly occupied orbitals, but is a reasonable approximation to the fluorine $2p$ orbital energy. This means that, compared to the ionized states ($H^+ + e^-$, and $F^+ + e^-$) of these atoms which may be taken as having zero energy, the fluorine atom is $17·42 - 13·6 = 3·82$ ev more stable than the hydrogen atom.

The relative energy levels of the $1s_H$ and $2p_F$ orbitals are shown in Figure 7.17 together with the bonding and antibonding molecular orbital energies. The formation of the bonding molecular orbital is shown in Figure 7.18.

It is of interest to enquire into the consequences of the participation of orbitals of different energies in the molecular orbital system. The bonding molecular orbital is much closer in energy to the original $2p_z$ orbital of the fluorine atom than it is to the $1s$ orbital of the hydrogen atom. It would, therefore, be expected that the sigma bonding orbital would resemble the $2p$ orbital more than it would resemble the $1s$ orbital. As can be seen from Figure 7.18, the bonding orbital does have two lobes as does the $2p$ orbital.

The hydrogen fluoride molecule possesses a *dipole moment*. Viewed as two point charges, $+\delta e$ and $-\delta e$, separated by a distance r, the dipole moment is given by

$$D = \delta e \times r \tag{7.49}$$

and is zero for *homonuclear* diatomic molecules.

The dipole moment of the hydrogen fluoride molecule is observed to have the value $1·91 \times 10^{-18}$ e.s.u. The internuclear distance is 0·092 nm so that the magnitude of the charge separation in the molecule is

$$\delta = \frac{1·91 \times 10^{-18}}{0·92 \times 10^{-8} \times 4·8 \times 10^{-10}} = 0·432 \tag{7.50}$$

Combination of Atoms: The Formation of Diatomic Molecules

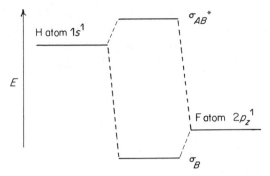

Figure 7.17. Approximate energies of the 1s orbital of the hydrogen and the $2p_z$ orbital of the fluorine atom

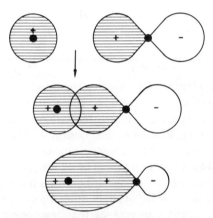

Figure 7.18. The formation of a bonding molecular orbital from the symmetric combination of an s and a p orbital

The molecule may be represented as

$$+0.43\,e \qquad -0.43\,e$$

$$\text{H} \longleftarrow 0.092\text{ nm} \longrightarrow \text{F}$$

with the fluorine atom being negative with respect to the hydrogen atom. This would be expected from the greater contribution of the fluorine orbital to the bonding orbital. In the hydrogen fluoride molecule the two electrons existing in the bonding orbital will not be symmetrically disposed between the two nuclei. The bonding orbital is, under these circumstances, better represented as in Figure 7.19, which shows a build-up of electron probability in the region of the fluorine atom.

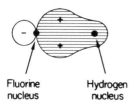

Figure 7.19. The bonding molecular orbital of HF showing the unsymmetrical sharing of electron probability between the two nuclei

The wave-mechanical description of this situation is to write the linear combinations of atomic orbitals as:

$$\psi_{\text{bonding}} = \psi_{1s}(\text{H}) + \lambda\psi_{2p}(\text{F}) \qquad (7.51)$$

and

$$\psi_{\text{antibonding}} = \lambda\psi_{1s}(\text{H}) - \psi_{2p}(\text{F}) \qquad (7.52)$$

the factor λ being included to allow for the unequal contributions of the atomic orbitals to the molecular ones. In equations (7.51) and (7.52) the value of λ would be expected to be greater than one since the fluorine orbital makes a major contribution to the bonding orbital and a minor one to the antibonding orbital. For a homonuclear diatomic molecule the coefficient λ would have a value of unity. (The equations 7.51 and 7.52 are not normalized.) The value of the coefficient deviates from unity in any system where the combining atomic orbitals differ in energy. It is possible to improve the quantitative aspects of the description of the hydrogen fluoride molecule by using an *sp* hybrid fluorine atomic orbital instead of a pure 2*p* atomic orbital. In this case the other *sp* hybrid orbital would accommodate a lone pair of electrons.

7.7 Electronegativity Coefficients

The asymmetric electron distribution in the molecule HF leads to the concept of electronegativity of atoms. Electronegativity is best stated to be the electron attracting power possessed by atoms when they are in a state of combination. Consider the hypothetical case of the combination of two unlike atoms, *A* and *B*, in the gas phase. There are two extreme products of such a reaction which are produced by transferring an electron from atom *A* to atom *B* or vice versa. These reactions may be written:

Combination of Atoms: The Formation of Diatomic Molecules

$$A_{(g)} + B_{(g)} \rightarrow A^+_{(g)} + B^-_{(g)} \tag{7.53}$$

$$A_{(g)} + B_{(g)} \rightarrow A^-_{(g)} + B^+_{(g)} \tag{7.54}$$

The energy changes occurring in these reactions may be calculated. In reaction (7.53) an electron has to be removed from atom A which will require an amount of energy equivalent to the ionization potential of the atom, I_A. When the electron is placed on atom B an amount of energy known as the electron affinity of atom B, E_B, is *released*. The total energy change in reaction (7.53) is given by

$$\Delta E_1 = I_A - E_B \tag{7.55}$$

By similar arguments the energy change occurring in reaction (7.54) is given by

$$\Delta E_2 = I_B - E_A \tag{7.56}$$

Which of these reactions occurs depends on the relative electronegativities of the atoms A and B. If B is the more electronegative (as we shall assume) then reaction (7.53) will be the preferred reaction and ΔE_1 will be smaller than ΔE_2, i.e.

$$\Delta E_1 < \Delta E_2 \tag{7.57}$$

which may be written in the form:

$$I_A - E_B < I_B - E_A \tag{7.58}$$

which inequality, upon rearrangement of terms in A to the left and terms in B to the right, becomes

$$I_A + E_A < I_B + E_B \tag{7.59}$$

So, starting with the assumption that $x_A < x_B$, where x represents electronegativity coefficients, we find the same inequality holding for the sums of ionization potentials and electron affinities for these elements. Mulliken suggested that the arithmetic mean of the quantities I and E would be a reasonable measure of the electronegativity of an element. This has proved to be so for the elements for which the calculation could be completed accurately, the ionization potentials of the elements being measurable with great accuracy but the electron affinities being somewhat subject to error. The lack of good electron affinity data has restricted the scope of the assignment of an electronegativity to the elements. Some values are given in Table 7.2. Pauling was responsible for the first scale of electronegativities which is based on the postulate that $\Delta_{AB}^{1/2}$ is proportional to

Table 7.2. Mulliken and Pauling Electronegativities for Some Elements

Element	Mulliken coefficient (x_M)	Pauling coefficient (x)
H	7·17	2·1
Li	2·96	—
C	5·61	2·5
N	7·34	3·0
O	9·99	3·5
Br	8·70	2·8
Cl	9·45	3·0
F	12·32	4·0

$x_A - x_B$, where x again represents the electronegativity coefficients of the elements A and B, and Δ is the difference between the actual bond dissociation energy of the $A-B$ bond (D_{AB}) and its estimated energy for 100 per cent covalent character, D_{cov}. This latter quantity was calculated as being the geometric mean* of the bond dissociation energies of the molecules A_2 and B_2 (D_{A_2} and D_{B_2}) so that

$$D_{cov} = ([D_{A_2}] \cdot [D_{B_2}])^{1/2} \tag{7.60}$$

and

$$\Delta_{AB} = D_{AB} - D_{cov} \tag{7.61}$$

all energy values being expressed in electron volts.

This scale is again lacking in data for D_{A_2} and D_{B_2} as well as having been based on the calculation of D_{cov} (equation 7.60) which is without theoretical justification. However, as can be seen from Table 7.2 which lists some values for Pauling electronegativity coefficients, they are in the same general order as those derived by Mulliken.

Of the more recent attempts to construct a universal scale of electronegativity, the most successful has been that due to Allred and Rochow who used the definition for electronegativity as 'the force of attraction between an atom and an electron, separated from the nucleus of that atom by a distance equal to the covalent radius of the atom'. The covalent radius of an atom is half the internuclear distance in a diatomic molecule formed from two such atoms. This is known for the majority of elements. The force of attraction referred to in the definition is given by the simple coulombic expression:

$$F = \frac{Z_{eff} \, e^2}{r_{cov}^2} \tag{7.62}$$

* In an earlier derivation the arithmetic mean was used.

Combination of Atoms: The Formation of Diatomic Molecules

where Z_{eff} is the effective nuclear charge, calculated using Slater's Rules (Appendix I). Allred and Rochow found that the empirical formula,

$$x = \frac{0 \cdot 359 Z_{eff}^2}{r_{cov}^2} + 0 \cdot 744 \tag{7.63}$$

reproduced the Pauling values for electronegativity coefficients and may then be applied to the rest of the elements for which Pauling and Mulliken values were not available. The periodic table shown in Figure 6.10 has the Allred and Rochow values for electronegativity coefficients under each element symbol (bottom right-hand corner).

There are two generalizations of importance to the overall understanding of the chemistry of the elements. These are that in descending the main groups of the periodic table there is a general decrease in electronegativity, while traversing the periods there is a general increase in electronegativity. This means that the periodic table is divided into a region of highly electropositive elements at the left-hand side—Groups I and II—with the transition series separating them from the more electronegative elements on the right-hand side.

7.8 Some Other Heteronuclear Diatomic Molecules

Molecular orbital theory may be applied successfully to other heteronuclear diatomic molecules, the discussion being restricted here to nitric oxide, NO, and carbon monoxide, CO.

Nitric Oxide

The electronic configurations of the nitrogen and oxygen molecules have been derived in Section 7.4. They are repeated here:

$$N_2 \; KK(\sigma_{2s})^2 \, (\sigma_{2s}^*)^2 \, (\pi_{2p_x})^2 \, (\pi_{2p_y})^2 \, (\sigma_{2p_z})^2$$

$$O_2 \; KK(\sigma_{2s})^2 \, (\sigma_{2s}^*)^2 \, (\sigma_{2p_z})^2 \, (\pi_{2p_x})^2 \, (\pi_{2p_y})^2 \, (\pi_{2p_x}^*)^1 \, (\pi_{2p_y}^*)^1$$

The bond orders in the N_2 and O_2 molecules are 3 and 2 respectively, and these values correlate with the observed lengths and strengths of the bonds.

The nitric oxide molecule contains fifteen electrons which must occupy the appropriate molecular orbital system. Its electronic configuration is

$$NO \; KK(\sigma_{2s})^2 \, (\sigma_{2s}^*)^2 \, (\sigma_{2p_z})^2 \, (\pi_{2p_x})^2 \, (\pi_{2p_y})^2 \, (\pi_{2p_x}^*)^1$$

with one electron occupying the doubly degenerate antibonding π_{2p} level. The bond order is 2·5 (the mean of those of N_2 and O_2), and the length and strength of the bond are consistent with this prediction from theory. The necessary data are shown in Table 7.3.

Table 7.3. Bond Length and Strength Data for the N_2, O_2 and NO Molecules

Molecule	Bond order	Bond length (nm)	Bond dissociation energy	
			(kcal.mole^{-1})	(kJ.mole^{-1})
N_2	3·0	0·109	225	940
NO	2·5	0·114	150	626
O_2	2·0	0·121	118	493

The relationships between the lengths and strengths of the bonds of the three molecules and the predicted bond orders is shown graphically in Figure 7.20. The relative energies of the higher orbitals of the three molecules have been plotted in Figure 2.16, and it may be seen from this that the σ_{2p_z} and π_{2p} orbital energies change from N_2 to O_2, and that the order of filling in nitric oxide is the same as in the oxygen molecule.

As may be seen from Table 7.2, oxygen is more electronegative than nitrogen which means that the oxygen orbitals will contribute more to the bonding orbitals than will the nitrogen orbitals. The reverse will be

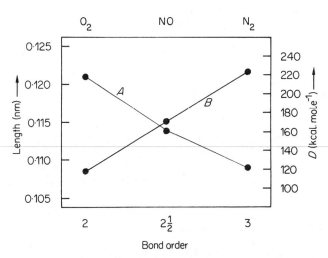

Figure 7.20. The bond lengths (*A*) and bond dissociation energies (*B*) of the molecules, O_2, NO and N_2, plotted against their respective bond orders

Combination of Atoms: The Formation of Diatomic Molecules 141

true for the antibonding orbitals. Since the orbitals combining to form the molecular system do so in such a way that each pair has the same atomic label (e.g. $2s$, $2p_x$, etc.), it is still possible to label the molecular orbitals in the same manner. It is not possible to do this in the case of the carbon monoxide molecule.

Carbon Monoxide

This molecule contains fourteen electrons which in the ground state occupy the seven lowest molecular orbitals. In combining two atomic orbitals to form two molecular ones it has already been pointed out (Section 7.5) that the energies of the two orbitals must not be too different. The first ionization potentials of the carbon and oxygen atoms (11·3 and 13·6 ev respectively) are reasonably close together but the $2p$–$2s$ energy gap is greater in the oxygen atom (16 ev) than it is in the carbon atom (4·5 ev). In consequence it is not possible for the $2s$ orbitals of the carbon and oxygen atoms to combine to give a σ–σ^* pair of molecular orbitals. The very low energy $2s$ orbital of the oxygen atom remains in the molecule as a lone-pair orbital. The $2s$ orbital of the carbon atom is sufficiently close in energy to the $2p$ levels to participate in sp hybridization (see Figure 7.15). The sp hybrid orbital directed towards the oxygen atom overlaps with, say, the $2p_z$ orbital of that atom to produce two sigma molecular orbitals. These two orbitals may not be given a simple description since they are formed from unlike atomic orbitals.

The molecular orbital scheme for carbon monoxide is summarized in Table 7.4, in which the orbitals appear in order of decreasing energy. A simple form of labelling of the orbitals (suggested by Mulliken) is also

Table 7.4. The Molecular Orbitals of the Carbon Monoxide Molecule

Molecular orbital	Carbon orbital	Oxygen orbital	Description	Label
8	sp_1	$2p_z$	Antibonding (σ^*)	u
6, 7	$2p_x, 2p_y$	$2p_x, 2p_y$	Antibonding (π^*)	v
5	sp_2	—	Non-bonding (σ)	w
3, 4	$2p_x, 2p_y$	$2p_x, 2p_y$	Bonding (π)	x
2	sp_1	$2p_z$	Bonding (σ)	y
1	—	$2s$	Non-bonding (σ)	z

shown where the lowest energy orbital is called the z orbital and in this case it has sigma character. The orbitals of higher energies are labelled according to reverse alphabetical order, together with their orbital characters. The fourteen electrons of the carbon monoxide molecule will thus have the configuration:

$$\text{CO } KK(z\sigma)^2 \, (y\sigma)^2 \, (x\pi)^4 \, (w\sigma)^2$$

The $x\pi$ and $y\sigma$ orbitals are bonding and the $z\sigma$ and $w\sigma$ are non-bonding in character. The bond order is therefore three, and a short, strong bond is expected. The data for this bond and for other bonds between carbon and oxygen atoms are shown in Table 7.5.

Table 7.5. Bond Strength and Length Data for Bonds Between Carbon and Oxygen Atoms

Molecule	Bond order	Bond length (nm)	Carbon–oxygen bond energy	
			(kcal.mole^{-1})	(kJ.mole^{-1})
Carbon monoxide	3	0·113	256	1090
Acetone	2	0·123	169	706
Ether	1	0·137	82	342

A consequence of the use of sp hybrid orbitals by the carbon atom is that a lone-pair of electrons (in the $w\sigma$ orbital) will be directed away from the carbon atom along the molecular axis. Carbon monoxide can act as an electron pair *donor* as in the molecule Ni(CO)$_4$ where the nickel–carbon and carbon–oxygen bonds are colinear. The bonding in similar molecules is discussed in detail in Chapter 11.

The presence of electrons in the $w\sigma$ non-bonding orbital of carbon monoxide is evident from the lower ionization potential of the molecule (14 ev) as compared to that for the isoelectronic nitrogen molecule (15·58 ev) in which the highest energy electrons are in the weakly bonding σ_{2p_z} orbital. Nitrogen, N$_2$, shows little tendency to donate electrons. The $2p$–$2s$ energy gap in the nitrogen atom is 11 ev and is large enough to ensure that hybridization is of little importance in determining the molecular orbital scheme. Thus it is to be noted that isoelectronic molecules do not necessarily have identical electronic configurations which in turn lead to different chemical behaviour.

Combination of Atoms: The Formation of Diatomic Molecules

Problems

7.1 Show that the wave functions expressed by equations (7.2) and (7.3) are orthogonal, i.e. that

$$\int_0^\infty \psi_S \cdot \psi_{AS} \cdot d\tau = 0$$

7.2 Using the radial functions given in Figure 5.4 for the $2p$ orbital of a hydrogen atom, draw the overlap diagrams corresponding to the molecular orbitals represented by equations (7.38) and (7.39) with an internuclear separation of 0·074 nm.

7.3 Write the electronic configurations of O_2, NO and N_2 in terms of the Mulliken symbols used in Section 7.7.

7.4 Plot a correlation diagram of the bond lengths and bond orders of all the diatomic species (neutral, positive or negative) for which data are given in this chapter.

7.5 Predict the effects upon the bond order and length of the O–O bond when an electron is ionized from

(a) the $v\pi$ orbital,
(b) the $w\pi$ orbital,
(c) the $x\sigma$ orbital, and
(d) the $y\sigma$ orbital.

8

Triatomic Molecules

In this chapter largely qualitative treatments of several triatomic molecular systems are developed. These may be divided into two main approaches:
(1) the Sidgewick–Powell theory that the shape and bonding of a molecule is determined by the minimization of the electrostatic repulsions between the various pairs of electrons in the valency shell of the central atom. The principles of electron correlation (Section 6.2) are implicit in the Sidgewick–Powell scheme. One example will be given to illustrate the latter statement. In the water molecule, OH_2, the oxygen atom will have eight electrons in its valency shell (i.e. $2s$, $2p$ orbitals). The molecule as a whole is diamagnetic and this means that four of the electrons will have one spin value (α) and the other four will have the alternative spin value (β). Taking each set separately, charge and spin correlation effects will work in the same direction to ensure that, in each set of four, electrons will be as far away from each other as possible. Such a configuration would be that of two separate uncorrelated tetrahedral sets of electrons. Spin correlation would tend to allow the two separate spin sets to come together and *share* corners of a regular tetrahedron. It must be pointed out that such a coincidence of the two spin-sets is opposed by charge correlation, and is only feasible due to the effects of the two protons in the OH_2 molecule. Such electrostatic localization of the two bonding pairs also localized the non-bonding electrons into pairs. Whether the two spin-sets in the neon atom are coincident or otherwise is an open matter, since there are no protons about in that case to throw any electrostatic advantage to a coinciding spin-set state of four electron pairs. In other words, the opposing effects of charge and spin correlation are difficult to estimate for the majority of electronic configurations, but nevertheless it is extremely useful to think in terms of electron pairs even though these might not be a true description of the state of a molecule.

Triatomic Molecules

(2) the Walsh approach which correlates the molecular orbitals of the extreme possible shapes of the molecule.

Both approaches are used to discuss the shapes and bonding in two classes of triatomic molecules, these being:

(a) AH_2, hydrides of elements, A, using only s and p electrons in the bonding orbitals, and
(b) AB_2 (and BAC), other compounds of the element, A, in which the atoms B and C may use orbitals other than $1s$.

8.1 Triatomic Hydrides, AH_2

It is perhaps most satisfactory to begin this section with some facts about triatomic hydride molecules. In Table 8.1 are shown the experimentally determined bond angles for a selection of such hydrides. The asterisk indicates that the molecules are free radicals of transitory existence.

Most of the molecules in Table 8.1 are unstable free radicals, but a discussion of their bonding and shapes is of considerable importance in

Table 8.1. The Bond Angles of Some Triatomic Hydride Molecules

Molecule	Bond angle
BH_2*	131°
CH_2* (singlet state)	104°
NH_2*	103·3°
OH_2	104·5°
CH_2* (triplet state)	180°

its bearing upon the bonding and shapes of more complex (but stable) molecules. A stable molecule whose bonding and shape are relevant to those of triatomic molecules is mercury dimethyl, $Hg(CH_3)_2$, in which the two carbon atoms and the mercury atom are collinear. It is of interest to consider the three atom carbon–mercury–carbon combination.

The mercury atom has a ground state electronic configuration in which all the electrons are paired. The highest energy occupied orbital is the $6s$. In its ground state it is not possible for the mercury atom to form electron pair bonds by sharing electrons with other atoms. Combination with other atoms can usually only occur when the overlapping orbitals are singly occupied. A possible *divalent* state of the mercury atom may be produced by *promoting* one of the 6s electrons to the 6p level. The most stable $6s^1 6p^1$ spectroscopic state is the ^3P state as was shown in Section 6.1. To produce this state from the $^1S_0$ electronic ground state requires an expenditure of 112 kcal.mole^{-1} (510 kJ.mole^{-1}) of energy. The actual valence state is not an observable spectroscopic state and the atom is not necessarily in a triplet condition as in the ^3P state. This is to allow combination with the orbitals of the carbon atoms which contain one electron each whose spin values are not specified. Any randomization of electron spins will incur a further expenditure of energy since the ^3P state is the most stable $6s^1 6p^1$ state. The valence state contains a mixture of spectroscopic states and allows for all possible combinations of spin states with those of the two methyl groups. In compound formation two bonds are produced between the mercury atom and the two methyl groups and in consequence an amount of energy is released. This amount is greater than that required to produce the valence state from the ground state. The situation is summarized in Figure 8.1. It must be understood that the energy changes as indicated in Figure 8.1 *do not* necessarily represent the actual changes which occur in the formation of the combined state.

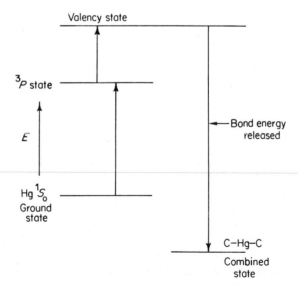

Figure 8.1. A representation of the various energy states of a mercury atom as used in the formation of mercury dimethyl

Triatomic Molecules

At this stage it is of extreme importance to link the valence state configuration of the mercury atom with the eventual shape of the carbon–mercury–carbon combination. The latter is known to be linear. To allow for this the valence state may be considered to be *sp* hybridized (Section 7.5), the two *sp* hybrid orbitals being directed away from each other so that the bond angle produced is 180°. The formation of the linear system of *sp* hybrid orbitals is shown in Figure 7.14.

The bonding situation in the hypothetical gaseous beryllium hydride molecule may be visualized in exactly the same way as that considered for mercury dimethyl. The $2s^2$ ground state ($^1S_0$) is converted to an upper state $2s^1 2p^1$ configuration (which may be the ^3P state) and this may in turn become the $sp(A)$, $sp(B)$ hybridized valence state. The use of the two *sp* hybrid orbitals in the formation of two localized bonds makes the linear bonding understandable. It is important to note that hybridization theory cannot of itself predict the shape of a molecule. It can only explain the shape (determined experimentally) in terms of the modification of the atomic orbitals of the central atom.

A simple approach to the prediction of molecular shape is due to Sidgewick and Powell. The basis of the method is to determine the number of electron pairs in the 'valence shell' of the central atom. In the cases of mercury dimethyl and the beryllium hydride molecules this number would be two. Considering that the electron pairs will repel each other the final shape of the molecule will be that in which such repulsion is minimized.

The principle of the minimization of repulsion between electron pairs is the basis of the Sidgewick–Powell method of predicting molecular shape. It is a principle of general application and has been extended to cover many different covalent systems by Linnett and by Nyholm and Gillespie.

The principle may now be applied to the molecules in Table 8.1.

BH_2

The central atom, boron, possesses the configuration, $2s^2 2p^1$, in its valence shell. The $1s^2$ pair of electrons are of too low an energy to be used in any chemical bonding. The boron atom may be converted to a *trivalent* state by an *s–p* promotion, this resulting in the $2s^1 2p^2$ configuration. Two of the singly occupied orbitals may be used to form the two localized sigma bonds, there being a singly occupied orbital which does not participate in bonding. The three centres of electronic charge, the two bonding pairs of electrons and the single non-bonding electron, then repel each other until a position of minimum repulsion is achieved. The position in this case would be that where the three regions of charge would be coplanar (with the central atom) and that the angle between the two bonding pairs would be the greatest of the three angles. The situation is shown in Figure 8.2.

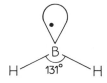

Figure 8.2. The distribution of two bonding pairs of electrons and the single non-bonding electron in the molecule BH_2

The Sidgewick–Powell idea would indicate that the $H\hat{B}H$ bond angle should be somewhat greater than 120°. The latter angle would be predicted for three equivalent centres of repulsion. In the BH_2 molecule one position is occupied by a single electron which must have a reduced repelling power compared with a bonding pair.

The above ideas are reasonably compatible with the observed bond angle of 131°.

CH_2 (*singlet and triplet states*)

The ground state of the carbon atom, $1s^2\ 2s^2\ 2p^2$, is a divalent state. If singly occupied orbitals are used to form two localized carbon–hydrogen sigma bonds then the resulting molecule, CH_2, would be expected to have a bond angle of 120°. Such a conclusion is based upon the fact that there would be three pairs of valence shell electrons surrounding the central carbon atom. These would repel each other, the most stable configuration being that in which the three electron pairs were directed towards the vertices of an equilateral triangle, in the centre of which would be the carbon atom. This situation is shown in Figure 8.3. That the angle is observed to be 104° for this singlet state (all electron spins paired so that $\Sigma s = 0$) could possibly be accounted for by considering that lone pairs of electrons have a greater repelling power than do

Figure 8.3. The distribution of the non-bonding and two bonding pairs of electrons in the molecule CH_2 in its singlet state

Triatomic Molecules

bonding pairs. This is logical since a lone pair would be localized on the central atom while a bonding pair would be distributed over the interatomic region of the bond. The lone pair represents a more concentrated source of repulsion and while we could expect lone pair–bond pair repulsions to be larger than bond pair–bond pair repulsions, it is not possible to predict the magnitude of the effect. It is thus possible to state with reasonable accuracy that the bond angle in singlet CH_2 is to be expected to be less than 120°, but it is not possible to say how much less.

That the angle in CH_2 is *not* 90° implies that the pure p atomic orbitals of the carbon atom are not responsible for the bonding. In order that the angle should be explained in terms of the overlapping of two orbitals of the carbon atom with the hydrogen orbitals it is necessary to postulate some suitable hybridization of the carbon orbitals.

A bond angle of 180° may be interpreted in terms of sp hybridization (Section 7.5), and these other s, p hybrid schemes are summarized in Table 8.2.

Table 8.2. The Hybridization of s and p Atomic Orbitals

Hybrid scheme	Number of atomic orbitals		Spatial orientation of hybrid orbitals	Bond angle
	(s)	(p)		
sp	1	1	Linear	180°
sp^2	1	2	Trigonally planar	120°
sp^3	1	3	Tetrahedral	109° 28'

These hybrid schemes are not the only ones which can be proposed but they are the limiting ones in which whole-number contributions from the atomic orbitals are involved.

In the triplet state of the CH_2 molecule there are two unpaired electrons with parallel spins ($\Sigma s = 1$), and this may only be achieved by making use of the potentially tetravalent state of the carbon atom. This latter state is achieved by an $s \rightarrow p$ promotion to give a $2s^1 2p^3$ configuration. Use of two of the singly occupied orbitals in bond formation leaves two singly occupied non-bonding orbitals. The Sidgewick–Powell approach would indicate that the four centres of negative charge would form a distorted tetrahedral arrangement in which the singly occupied orbitals would probably be closer together than the two bonding orbitals. The bond angle would be predicted to be of the order of 109° 28' or greater.

Again the hybridization of atomic carbon orbitals would have to be invoked to explain how the s and p orbitals could be rearranged to give such an angle. In this case the hybridization would be allied to sp^3 type.

The triplet state of CH_2 in fact has a bond angle of 180° and would seem not to fit in with the Sidgewick–Powell prediction.

OH_2

The ground state of the oxygen atom, (He) $2s^2\ 2p^4$, is divalent, and in forming two localized bonds as in the water molecule there would be four pairs of 'valency' electrons surrounding the central atom. These would repel each other to a tetrahedral minimum repulsion configuration. Two positions would be occupied by the bonding pairs and two by the lone pairs. As pointed out previously the lone pairs are a more concentrated source of repulsion, and it would be expected that the repulsion between the two lone pairs would be greater even than that between lone and bonding pairs. These unequal repulsion effects would result in the angle between the bonding pairs being less than the regular tetrahedral angle. This is so in the water molecule.

NH_2

This radical has one electron less than the water molecule and can be compared with the H_2O^+ ion with which it is isoelectronic. If an electron is removed from one of the lone pair orbitals in the water molecule (the bonding electrons are more stable) then one could predict an opening of the bond angle in terms of reduced repulsion of the bonding pairs by the non-bonding electrons. This does not happen since the bond angle in NH_2 is 0·9° *smaller* than that in water. The bond angle in H_2O^+ is not appreciably different from that in the neutral molecule so it would appear that the Sidgewick–Powell approach is lacking in this case.

8.2 The Walsh Treatment of Triatomic Hydrides: Symmetry Theory

The basis of the Walsh treatment of triatomic hydride molecules is to consider the nature and energies of the orbitals which may be used to describe the two extreme shapes of these molecules. The orbitals and their energies are then 'correlated' so that the variations of orbital energies with bond angle are shown. The resultant bond angle in a system may be decided by the number of electrons present and their placement in the orbitals which give a minimum total energy.

The two extreme shapes for a triatomic hydride are those in which the

Triatomic Molecules

bond angle is either 180° (linear) or 90° (bent). These are the shapes for which the orbitals used are easily understood.

In dealing with polyatomic molecules it is frequently useful to refer to the symmetry properties of the orbitals of the molecule as they compare with those of the molecule itself. It is not the purpose of this book to deal in detail with symmetry, but it is important at this stage to have an idea of its application. The application of symmetry (group theory) to simple polyatomic molecules forms an excellent example of its uses.

Elements of Symmetry

A molecule may possess several elements of symmetry which may include:

(1) axes of symmetry,
(2) planes of symmetry, and
(3) an inversion centre or centre of symmetry.

An *axis of symmetry* may be defined as an axis around which rotation of the molecule by a certain angle produces an equivalent configuration of the molecule. The water molecule possesses one such axis which lies in the molecular plane and bisects the $\hat{HOH}$ angle. This is shown in Figure 8.4. Rotation of the molecule around this axis by 180° produces an equivalent configuration of the molecule as is shown in Figure 8.5, where the hydrogen atoms are distinguished by subscripts.

The symmetry operation of rotation through 180° about the symmetry axis causes the hydrogen atoms to interchange positions without altering the configuration of the molecule in space. This particular *symmetry operation* is given the symbol C_2, this being also the symbol for an axis of symmetry of order *two*. The order of the symmetry axis is simply 360° divided by the number of degrees through which it is necessary to rotate

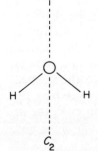

Figure 8.4. Showing the C_2 axis of symmetry which bisects the HOH angle of the water molecule and which is contained by the molecular plane

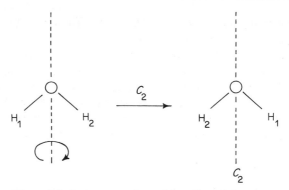

Figure 8.5. A representation of the effect of the operation C_2 (i.e. rotation through 180° about the C_2 axis) upon the positions of atoms of the water molecule

the molecule to produce an equivalent configuration (i.e. 180° in this case). The C is the symbol for an axis of symmetry.

A linear molecule possesses a C_∞ axis of symmetry which is coincident with the molecular axis. There are an infinite number of rotation operations about this axis which *all* produce equivalent configurations of the molecule.

A *plane of symmetry* may be defined as a plane through which reflexion of all the atoms of the molecule produces an equivalent configuration of that molecule. The water molecule possesses two planes of symmetry, the one being the molecular plane and the other is perpendicular to this, both planes containing the C_2 axis of symmetry and are known as vertical planes since they contain the C_2 axis which in this case is the only axis of symmetry. These planes of symmetry are shown in Figure 8.6.

A homonuclear diatomic molecule (e.g. H_2, O_2), as well as possessing an infinite number of vertical planes, has a horizontal plane which bisects the line joining the two nuclei and is *perpendicular* to the main axis of symmetry, this latter symmetry element (C_∞) coinciding with the molecular axis. The symbol for a symmetry plane is σ and if the plane is a vertical one this is indicated by a subscript, viz.: σ_v. In the case of the water molecule which possesses two σ_v, it is customary to differentiate them by having the molecular plane represented by σ'_v. The operation of reflexion of the water molecule in σ'_v is also represented by the symbol, σ'_v, and does not change the positions of the three atoms since they are all contained by the plane. The operation, σ_v (reflexion through the σ_v plane which bisects the molecule), leaves the oxygen atom alone and reverses the positions of the hydrogen atoms.

Triatomic Molecules

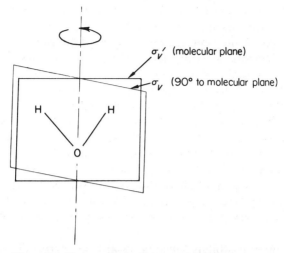

Figure 8.6. The two vertical planes of symmetry and the C_2 axis of symmetry of the water molecule

A *centre of symmetry* or an *inversion centre* may be defined as the point through which a molecule may be inverted so as to produce an equivalent configuration. It is symbolized by i, which letter also indicates the operation of inversion of the molecule.

A linear AH_2 molecule possesses an inversion centre which coincides with the centre of the atom A. The water molecule does not possess such an element of symmetry.

A regular octahedron (see Figure 9.1(d)) possesses an inversion centre, but a regular tetrahedron (see Figure 9.1(b)) does not.

A trivial but mathematically (group theoretically) necessary symmetry element which was not included in the above list is the *identity*. This is symbolized by E (or sometimes I), and the associated symmetry operation is leaving the molecule alone. Such an operation obviously produces an equivalent molecular configuration (*identical* in this case).

It is now possible to use these symmetry elements to classify the molecular shapes and molecular orbitals associated with AH_2 molecules.

(a) *The Orbitals of a Linear Triatomic Hydride Molecule*

The central atom, A, may be assumed to be one involving s and p atomic orbitals, and these in some way have to be coupled to the $1s$ orbitals of the two hydrogen atoms.

In order that symmetry theory may be applied to the orbitals of the molecule it is necessary first of all to classify the atomic orbitals in terms of symmetry elements and their associated operations.

The orbitals of the central atom may be dealt with first of all. The important elements of symmetry are the C_∞ axis and the inversion centre. The s orbital is symmetric to both operations (C_∞ and i) and is therefore classified as a σ_g orbital, the g subscript (gerade = even) indicating that the orbital is symmetric with respect to inversion.

The p_y orbital which may arbitrarily be considered to lie along the molecular axis, C_∞, is also a sigma orbital in that rotation about the axis does not alter its symmetry properties. The operation of inversion, i, does alter the symmetry of the p_y orbital. The two lobes are interchanged and the positive and negative regions exchange positions. The orbital may be said to be antisymmetric with respect to inversion and such a property is indicated by a subscript u to the symmetry symbol, viz.: σ_u (ungerade = odd).

The two other p orbitals which necessarily are perpendicular to the p_y orbital and to the molecular axis are classified as pi orbitals. This is because they are not symmetric with respect to the C_∞ operation. They are also antisymmetric to inversion so that they both are π_u orbitals.

The 1s orbitals of the two hydrogen atoms in the linear AH_2 molecule are not symmetry orbitals. Taken separately they do not have any of the properties which would allow their classification in terms of sigma or pi orbitals. For instance, the operation of inversion carried out upon one of the orbitals would merely transfer it to the position of the other hydrogen orbital. Such an occurrence is neither symmetric nor antisymmetric with respect to the operation, and prevents symmetry classification of the single orbital within the terms of the linear molecule.

To remove such limitations it is necessary to form two *group orbitals* from the isolated 1s atomic orbitals. This is done by taking suitable linear combinations of the two atomic eigenfunctions, viz.:

$$G_1 = \frac{1}{\sqrt{2}}[\psi_{1s}(1) + \psi_{1s}(2)] \tag{8.1}$$

and

$$G_2 = \frac{1}{\sqrt{2}}[\psi_{1s}(1) - \psi_{1s}(2)] \tag{8.2}$$

G_1 and G_2 are the normalized group eigenfunctions. They represent the symmetric and antisymmetric combinations of the two 1s orbitals involved in the AH_2 molecule. A pictorial representation of the two group orbitals is shown in Figure 8.7. The overlap integrals for these two group orbitals are necessarily rather small since the component atomic orbitals are separated from each other by two conventional A—H bond lengths.

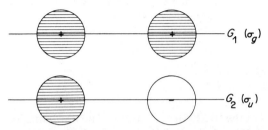

Figure 8.7. A representation of the two group orbitals produced by the symmetric (G_1) and antisymmetric (G_2) combinations of 1s atomic orbitals of the two hydrogen atoms in a linear AH_2 molecule

The group orbital, G_1, transforms with respect to the symmetry operations of a linear molecule as does a σ_g orbital. The antisymmetric combination, G_2, is a σ_u orbital.

The orbitals used in linear AH_2 molecules and their symmetry 'characters' are summarized in Table 8.3.

In order that these orbitals may be used to form a molecular system it is necessary to combine only those orbitals of the same character. In this case it is possible to combine the σ_g and σ_u orbitals of atom A with the corresponding G_1 and G_2 orbitals of the two hydrogen atoms. The π_u orbitals, therefore, must remain as non-bonding orbitals capable of accommodating lone (non-bonding) pairs of electrons.

The orbital combination of $\sigma_g(A)$ with $\sigma_g(G_1)$ in a symmetric manner

Table 8.3. Symmetry Characters of the Orbitals used in a Linear AH_2 Molecule

Orbital		Symmetry character
Central atom	s	σ_g
	p_x	π_u
	p_y	σ_u
	p_z	π_u
Hydrogen atoms	G_1	σ_g
	G_2	σ_u

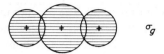

Figure 8.8. A representation of the symmetric combination of the G_1 group orbital with the s orbital of the atom A in a linear AH_2 molecule

is shown in Figure 8.8 and must not be mistaken for a combination of three complete orbitals. It is a combination of the σ_g orbital of the central atom with the symmetric group orbital, G_1. The latter is a 50–50 mixture of the s orbitals of the two hydrogen atoms. The normalized eigenfunction of the symmetric combination (bonding orbital) shown in Figure 8.8 is

$$\psi_1 = \frac{1}{\sqrt{2}}(\psi_s(A) + G_1) \tag{8.3}$$

That of the antisymmetric combination (antibonding orbital) is

$$\psi_5 = \frac{1}{\sqrt{2}}(\psi_s(A) - G_1) \tag{8.4}$$

which is represented in Figure 8.9. The normalized wave functions of the orbitals, ψ_1 and ψ_5, may be expanded using equations (8.1) and (8.2) and become:

$$\psi_1 = \frac{1}{\sqrt{2}}\psi_s(A) + \tfrac{1}{2}\psi_{1s}(1) + \tfrac{1}{2}\psi_{1s}(2) \tag{8.5}$$

and

$$\psi_5 = \frac{1}{\sqrt{2}}\psi_s(A) - \tfrac{1}{2}\psi_{1s}(1) - \tfrac{1}{2}\psi_{1s}(2) \tag{8.6}$$

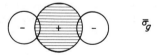

Figure 8.9. A representation of the antisymmetric combination of the G_1 group orbital with the s orbital of the atom A in a linear AH_2 molecule

Triatomic Molecules

The squares of the coefficients of the atomic eigenfunctions in equations such as (8.5) and (8.6) express the relative contributions of the orbitals to the molecular eigenfunction. When a molecular eigenfunction is normalized (the contributing atomic eigenfunctions being separately normalized) the squares of the coefficients of the atomic eigenfunctions express the *actual* contributions of the atomic orbitals to the molecular orbital. The squares of the coefficients in equations (8.5) and (8.6) are summarized in Table 8.4. From this it can be seen that the ratio of the atomic orbitals used to form molecular orbitals is as derived above and that the equivalent of *two* whole orbitals are used to form the bonding and antibonding combinations, ψ_1 and ψ_5, respectively. The formation of the orbitals, ψ_1 and ψ_5, is thus in accordance with the principle outlined in Section 7.3.

The other permitted combinations for the linear molecule are those which may be written as:

$$\psi_2 = \frac{1}{\sqrt{2}}(\psi_{p_y}(A) + G_2) \tag{8.7}$$

and

$$\psi_6 = \frac{1}{\sqrt{2}}(\psi_{p_y}(A) - G_2) \tag{8.8}$$

These orbitals are formed by combining symmetrically and antisymmetrically the σ_u orbital of the central atom, A, with the antisymmetric hydrogen group orbital, G_2, and are shown in Figure 8.10.

The equations (8.7) and (8.8) may be expanded to read:

$$\psi_2 = \frac{1}{\sqrt{2}}\psi_{p_y}(A) + \tfrac{1}{2}\psi_{1s}(1) - \tfrac{1}{2}\psi_{1s}(2) \tag{8.9}$$

and

$$\psi_6 = \frac{1}{\sqrt{2}}\psi_{p_y}(A) - \tfrac{1}{2}\psi_{1s}(1) + \tfrac{1}{2}\psi_{1s}(2) \tag{8.10}$$

Table 8.4. Orbital Contributions to ψ_1 and ψ_5

Atomic orbital	Square of coefficient in ψ_1 (equation 8.5)	ψ_5 (equation 8.6)	Totals
$\psi_s(A)$	$\tfrac{1}{2}$	$\tfrac{1}{2}$	1
$\psi_{1s}(1)$	$\tfrac{1}{4}$	$\tfrac{1}{4}$	$\tfrac{1}{2}$
$\psi_{1s}(2)$	$\tfrac{1}{4}$	$\tfrac{1}{4}$	$\tfrac{1}{2}$
Totals	1	1	2

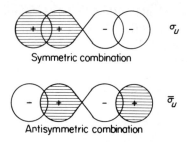

Figure 8.10. The symmetric and antisymmetric combinations of the G_2 group orbital with a p orbital of atom A in a linear AH_2 molecule

and as can be seen, the values of the coefficients of the contributing atomic wave functions are numerically the same as the corresponding ones in equations (8.5) and (8.6). Some of the signs are different, but this does not alter the squares of the coefficients.

This means that one p orbital and the equivalent of one s orbital are involved in producing the bonding (ψ_2) and antibonding (ψ_6) orbitals of this system.

The other two p orbitals of the central atom are π_u in character, and must remain as non-bonding orbitals. They are degenerate and their eigenfunctions may be written as:

$$\psi_3 = \psi_4 = \psi_p(A) \tag{8.11}$$

A summary of the bonding, non-bonding or antibonding characters of the six molecular orbitals of the linear AH_2 molecule is shown in Table 8.5. The character as far as A—H bonding is shown, and in addition the interaction of the hydrogen orbitals is shown. Since group orbitals are being used in this treatment it is possible to ascribe to them characters of bond-

Table 8.5. Bonding Characters of the Orbitals in a Linear AH_2 Molecule

	Orbital	Symmetry	A–H	H–H
↑	ψ_6	$\overline{\sigma_u}$ [a]	Antibonding	Antibonding
	ψ_5	$\overline{\sigma_g}$	Antibonding	Bonding
Orbital	ψ_3, ψ_4	π_u	Non-bonding	Non-bonding
energy	ψ_2	σ_u	Bonding	Antibonding
↓	ψ_1	σ_g	Bonding	Bonding

[a] The bars over the symmetry symbols indicate the antibonding (A–H) nature of the orbitals.

Triatomic Molecules

ing, non-bonding or antibonding with respect to atoms which are not normally thought of as being directly linked. Any effect so operating will naturally be of a smaller order of magnitude as compared to the effects operating between two adjacent atoms in a molecule.

The energies of the orbitals ψ_1 to ψ_6 may now be considered. The main difference in the energies of ψ_1 and ψ_2 is that the central atom uses an s orbital in ψ_1 but uses a higher energy p orbital (Section 5.5) in ψ_2. So that even if both orbitals are bonding with respect to the A—H interaction, electrons in ψ_1 will be more stable than those in ψ_2. Of the two orbitals which are bonding with respect to the A—H regions, i.e. ψ_1 and ψ_2, the former is placed at a lower energy than the latter, since the former orbital is also bonding as far as the hydrogen–hydrogen interaction is concerned. Similar reasoning allows ψ_5 to be of lower energy than ψ_6.

The relative energies of the six orbitals are shown in Figure 8.11, with the energies of the ψ_3, ψ_4 degenerate pair as the reference point.

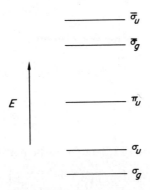

Figure 8.11. The relative energies of the orbitals ψ_1 to ψ_6 of a linear AH$_2$ molecule (qualitative)

One further point remains to be discussed as far as the linear AH$_2$ molecule is concerned, this being the reconciliation of the molecular orbital scheme just outlined with the more usual sp hybrid scheme of localized bonds.

Let us consider the bonding orbitals, ψ_1 and ψ_2. It is possible to make normalized linear combinations of them such that

$$K_1 = \frac{1}{\sqrt{2}}(\psi_1 + \psi_2) \tag{8.12}$$

and

$$K_2 = \frac{1}{\sqrt{2}}(\psi_1 - \psi_2) \qquad (8.13)$$

and these, upon expansion using equations (8.5) and (8.9) become:

$$K_1 = \frac{1}{\sqrt{2}}\left[\frac{1}{\sqrt{2}}\psi_s(A) + \tfrac{1}{2}\psi_{1s}(1) + \tfrac{1}{2}\psi_{1s}(2)\right.$$
$$\left. + \frac{1}{\sqrt{2}}\psi_{p_y}(A) + \tfrac{1}{2}\psi_{1s}(1) - \tfrac{1}{2}\psi_{1s}(2)\right]$$
$$= \tfrac{1}{2}\psi_s(A) + \tfrac{1}{2}\psi_{p_y}(A) + \frac{1}{\sqrt{2}}\psi_{1s}(1) \qquad (8.14)$$

and

$$K_2 = \frac{1}{\sqrt{2}}\left[\frac{1}{\sqrt{2}}\psi_s(A) + \tfrac{1}{2}\psi_{1s}(1) + \tfrac{1}{2}\psi_{1s}(2)\right.$$
$$\left. - \frac{1}{\sqrt{2}}\psi_{p_y}(A) - \tfrac{1}{2}\psi_{1s}(1) + \tfrac{1}{2}\psi_{1s}(2)\right]$$
$$= \tfrac{1}{2}\psi_s(A) - \tfrac{1}{2}\psi_{p_y}(A) + \frac{1}{\sqrt{2}}\psi_{1s}(2) \qquad (8.15)$$

The equations (8.14) and (8.15) may now be rearranged to read:

$$K_1 = \frac{1}{\sqrt{2}}\left[\frac{1}{\sqrt{2}}\psi_s(A) + \frac{1}{\sqrt{2}}\psi_{p_y}(A)\right] + \frac{1}{\sqrt{2}}\psi_{1s}(1) \qquad (8.16)$$

and

$$K_2 = \frac{1}{\sqrt{2}}\left[\frac{1}{\sqrt{2}}\psi_s(A) - \frac{1}{\sqrt{2}}\psi_{p_y}(A)\right] + \frac{1}{\sqrt{2}}\psi_{1s}(2) \qquad (8.17)$$

The terms in parentheses in equations (8.16) and (8.17) represent the normalized linear combinations of the s and the p_y orbitals of atom A. They should be compared with equations (7.45) and (7.46) in Section 7.5. Equations (8.16) and (8.17) are then seen to represent the bonding combinations of the two sp hybrid orbitals of atom A with the $1s$ atomic orbitals of the two hydrogen atoms (1) and (2) respectively.

The description of the bonding in AH_2 molecules by two localized bonds is therefore to be seen to be equivalent to the more complicated

Triatomic Molecules

three-centre orbital treatment which involves the use of group orbitals. The latter treatment, however, gives a better understanding of the available energy levels in such a molecule.

(b) *The Orbitals of a 90° Triatomic Hydride Molecule*

As has been pointed out in Section 8.1 a molecule, depending upon its shape, may possess one or more elements of symmetry. A triangular molecule such as AH_2 possesses four such symmetry elements. These are:

(1) a two-fold axis of symmetry, C_2, which bisects the $H\hat{A}H$ angle,
(2) a molecular plane of symmetry, σ'_v (the yz plane),
(3) a vertical plane of symmetry, σ_v (the xz plane), and
(4) the identity element, E.

It is possible to classify the orbitals of a bent molecule in terms of whether they are symmetric or antisymmetric to the four symmetry operations associated with the elements of symmetry of the molecule.

There are, in fact, only four different classes of orbital possible which are independent of each other. These are given in Table 8.6 which is the so-called character table for the C_{2v} point group, this latter being a summary in symmetry terms of the shape of a bent triatomic molecule.

Table 8.6. The C_{2v} Character Table

Symmetry symbol	E	C_2	$\sigma_v(xz)$	$\sigma'_v(yz)$
a_1	+1	+1	+1	+1
a_2	+1	+1	−1	−1
b_1	+1	−1	+1	−1
b_2	+1	−1	−1	+1

Table 8.6 shows the way in which the four classes of orbital or representations transform with respect to the four symmetry operations, E, C_2, σ_v and σ'_v. In the table +1 appears when the orbital is symmetric to the operation, and −1 when the orbital is antisymmetric to the operation.

An a_1 orbital is symmetric to all four operations, but the a_2 orbital is antisymmetric to the σ_v and σ'_v operations. The orbitals which are antisymmetric to the C_2 rotation are b orbitals. The two are distinguished by the subscripts, the b_2 being antisymmetric to the σ_v operation (as would be an a_2 orbital). The above represent all the possible representations which are independent. There are others which are merely combinations of these four.

It now remains for us to classify the actual orbitals in bent AH_2 molecules in these terms.

First of all we may make some arbitrary choices of the coordinates of the molecular system. It is usual to choose the C_2 axis to be coincident with the z coordinate axis, and to make the molecular plane the yz plane. The vertical plane is thus the xz plane. The AH_2 molecule and the three coordinate axes are shown in Figure 8.12.

With the molecule arranged as in Figure 8.12 and with the atomic orbitals of the central atom arranged normally, they may be classified in the following manner.

The s orbital is symmetric to all the operations and is designated a_1. The p_z orbital lies along the C_2 axis and is also symmetric to all the operations, and is thus a_1. The p_x orbital is symmetric to E and σ_v but antisymmetric to C_2 and σ'_v, and is therefore a b_1 orbital. The p_y orbital is symmetric to E and σ'_v but antisymmetric to C_2 and σ_v, and is therefore a b_2 orbital.

The 1s orbitals of the two hydrogen atoms may be combined to form group orbitals, G_1 and G_2, as represented by equations (8.1) and (8.2). The group orbital, G_1, transforms as an a_1 orbital, it being symmetric to all four operations. The G_2 orbital is symmetric to E and σ'_v but antisymmetric to C_2 and σ_v, and is therefore a b_2 orbital. In this particular molecular system no orbital is present which transforms as a_2.

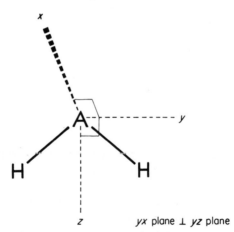

Figure 8.12. The position of an AH_2 molecule with respect to the x, y and z coordinates. The choice of position is arbitrary, the usual convention being to make the main symmetry axis (C_2) coincide with the z axis

Triatomic Molecules

A summary of the orbital symmetries for the bent AH_2 molecule is shown in Table 8.7.

Table 8.7. The Symmetry Characters of Atomic Orbitals used in a Bent AH_2 Molecule

Orbital		Symmetry character
Central atom	s	a_1
	p_x	b_1
	p_y	b_2
	p_z	a_1
Hydrogen atoms	G_1	a_1
	G_2	b_2

The hydrogen group orbitals, G_1 and G_2, of orbital symmetries a_1 and b_2 respectively can only combine with orbitals of the central atom of corresponding symmetries. The p_x orbital, therefore, is barred from participating in any combination, and is thus a non-bonding orbital.

Considering that the $s(a_1)$ orbital may be also of non-bonding character, it is possible to build two bonding–antibonding combinations from the remaining orbitals.

The symmetric group orbital, G_1, may be combined with the p_z orbital of the central atom to give bonding and antibonding orbitals:

$$\psi'_1 = \frac{1}{\sqrt{2}}(\psi_{p_z}(A) + G_1) \tag{8.18}$$

and

$$\psi'_5 = \frac{1}{\sqrt{2}}(\psi_{p_z}(A) - G_1) \tag{8.19}$$

The pictorial representations of these orbitals are shown in Figure 8.13.

The other possible combinations are between the G_2 group orbital and the p_y orbital of atom A. The eigenfunctions may be written as:

$$\psi'_2 = \frac{1}{\sqrt{2}}(\psi_{p_y}(A) + G_2) \tag{8.20}$$

and

$$\psi'_6 = \frac{1}{\sqrt{2}}(\psi_{p_y}(A) - G_2) \tag{8.21}$$

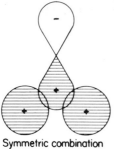

Symmetric combination
a_1

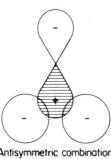

Antisymmetric combination
$\bar{a}_1$

Figure 8.13. The symmetric and antisymmetric combinations of the group orbital G_1 with the p_z orbital of atom A in a 90° AH_2 molecule

The pictorial representations of these orbitals are shown in Figure 8.14. The non-bonding orbitals, $s(a_1)$ and $p_x(b_2)$, may be written as:

$$\psi'_3 = \psi_s(A) \tag{8.22}$$

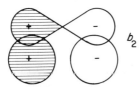

Symmetric combination

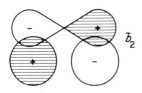

Antisymmetric combination

Figure 8.14. The symmetric and antisymmetric combinations of the group orbital G_2 with the p_y orbital of atom A in a 90° AH_2 molecule

Triatomic Molecules

and

$$\psi'_4 = \psi_{p_x}(A) \tag{8.23}$$

The orbitals and their symmetries and bonding characteristics are shown in Table 8.8.

Table 8.8. Bonding Characters of the Orbitals in a 90° AH_2 Molecule

Orbital	Symmetry	Character	
		A—H	H—H
ψ'_6	$\overline{b_2}$[a]	Antibonding	Antibonding
ψ'_5	$\overline{a_1}$	Antibonding	Bonding
ψ'_4	b_1	Non-bonding	Non-bonding
ψ'_3	a_1	Non-bonding	Non-bonding
ψ'_2	b_2	Bonding	Antibonding
ψ'_1	a_1	Bonding	Bonding

[a] The bars over the symmetry symbols indicate the antibonding (A–H) nature of the orbitals.

The energies of the orbitals of the bent molecule are generally in the same order as those of the linear molecules, save for the $\psi'_3 (a_1)$ orbital. This, in the absence of hybridization, is a non-bonding orbital of s character. Its energy is to be expected to be of the order of those of the bonding levels, ψ'_1 and ψ'_2, which involve only the p character from the central atom. A possible order of energies is shown in Figure 8.15.

The molecular orbitals in the bent AH_2 (90°) molecule as well as those

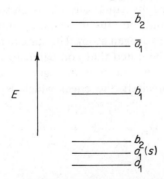

Figure 8.15. A possible order of the orbitals of a bent (90°) AH_2 molecule

of the linear molecule do not obviously make use of localized bonding. In the bent case two localized bonding orbitals may be made by taking linear combinations of the delocalized molecular bonding orbitals, viz.:

$$L_1 = \frac{1}{\sqrt{2}}(\psi'_1 + \psi'_2)$$

$$= \frac{1}{\sqrt{2}}\left[\frac{1}{\sqrt{2}}\psi_{p_z}(A) + \tfrac{1}{2}\psi_{1s}(1) + \tfrac{1}{2}\psi_{1s}(2)\right.$$

$$\left. + \frac{1}{\sqrt{2}}\psi_{p_y}(A) + \tfrac{1}{2}\psi_{1s}(1) - \tfrac{1}{2}\psi_{1s}(2)\right]$$

$$= \frac{1}{\sqrt{2}}\left[\frac{1}{\sqrt{2}}\psi_{p_z}(A) + \frac{1}{\sqrt{2}}\psi_{p_y}(A)\right] + \frac{1}{\sqrt{2}}\psi_{1s}(1) \qquad (8.24)$$

and

$$L_2 = \frac{1}{\sqrt{2}}(\psi'_1 - \psi'_2)$$

$$= \frac{1}{\sqrt{2}}\left[\frac{1}{\sqrt{2}}\psi_{p_z}(A) + \tfrac{1}{2}\psi_{1s}(1) + \tfrac{1}{2}\psi_{1s}(2)\right.$$

$$\left. - \frac{1}{\sqrt{2}}\psi_{p_y}(A) - \tfrac{1}{2}\psi_{1s}(1) + \tfrac{1}{2}\psi_{1s}(2)\right]$$

$$= \frac{1}{\sqrt{2}}\left[\frac{1}{\sqrt{2}}\psi_{p_z}(A) - \frac{1}{\sqrt{2}}\psi_{p_y}(A)\right] + \frac{1}{\sqrt{2}}\psi_{1s}(2) \qquad (8.25)$$

The orbitals represented by equations (8.24) and (8.25) are localized orbitals using pure p contributions from the central atom and the unmixed $1s$ orbitals of the hydrogen atom. The state of affairs is represented in Figure 8.16. It is to be noted that compared to their orientations in Figures 8.13 and 8.14, the p orbitals in Figure 8.16 have been moved through 45°. This is precisely the effect which is represented by the transformations,

$$\psi_{p_1} = \frac{1}{\sqrt{2}}\psi_{p_z} + \frac{1}{\sqrt{2}}\psi_{p_y} \qquad (8.26)$$

and

$$\psi_{p_2} = \frac{1}{\sqrt{2}}\psi_{p_z} - \frac{1}{\sqrt{2}}\psi_{p_y} \qquad (8.27)$$

where the 'new' p orbitals, ψ_{p_1} and ψ_{p_2}, are mutually perpendicular but

Triatomic Molecules

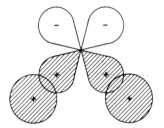

Figure 8.16. A representation of two localized bonding combinations in a bent (90°) AH_2 molecule

are rotated through an angle of 45° with respect to the original coordinate axes.

(c) *Correlation of the Energies of the Orbitals of Linear and 90° Bent AH_2 Molecules*

The relative energies of the symmetry orbitals of the linear and 90° bent AH_2 molecules have been shown in Figures 8.11 and 8.15 respectively.

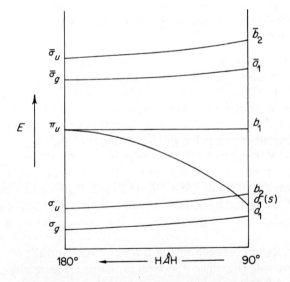

Figure 8.17. A Walsh Diagram for AH_2 molecules. The orbital energies are plotted as a function of the bond angle. (After Walsh, *J. Chem. Soc.*, **1953**, 2262)

These may be combined into the same energy diagram as is shown in Figure 8.17. The common reference energy level is the non-bonding level. In the linear state this is designated π_u, and in the bent state it is the b_1 level. The p_x orbital is non-bonding in both conformations and its energy is unaffected by bond angle. Thus a horizontal line links the π_u and b_1 levels in Figure 8.17. Other correlations may be made with ease. The two bonding levels may be joined up, i.e. the σ_g with the a_1 and the σ_u with the b_2.

The corresponding antibonding levels may also be 'correlated', the $\overline{\sigma_g}$ with the $\overline{a_1}$ and the $\overline{\sigma_u}$ with the $\overline{b_2}$ orbital. The only remaining correlation is between the second π_u orbital (it is a doubly degenerate level) and the $a_1(s)$ non-bonding orbital.

There are two important considerations in deciding upon the drawing of the correlations in the *Walsh Diagram* of Figure 8.17 which may be known as Walsh's Rules. These are:

(1) The energy of an orbital increases with increasing p character and decreasing s character. Thus the b_2 and a_1 orbitals are higher in energy respectively than the σ_u and σ_g orbitals which are of greater s character. The first rule is responsible for the main differences in orbital energies as the bond angle changes.

(2) If the molecular orbital is *antibonding* between the hydrogen atoms then a *linear* state is favoured, but if the molecular orbital is bonding with respect to the hydrogen atoms then a bent state is favoured.

For instance, the $\sigma_g - a_1$ orbital is bonding and, as far as Rule 2 is concerned, is more stable in the bent state where the hydrogen atoms are closer together than in the linear molecule. Rule 1, of course, indicates that, as the molecule bends, the energy of the $\sigma_g - a_1$ orbital will increase because of the lesser s and greater p contribution from the central atom. This effect is offset somewhat by the effect due to Rule 2.

The $\sigma_u - b_2$ bonding orbital (A—H) is H—H antibonding, and the application of the two rules indicates that the orbital energy will increase with the bending of the molecule due to Rule 2. Rule 1 is not relevant here since both the σ_u and b_2 orbitals are built from a p orbital of the central atom. Similar reasoning applies to the corresponding antibonding orbitals.

The correlation between one of the π_u orbitals and the b_1 orbital has been mentioned. The other π_u orbital (p character) correlates in the bent molecule with the a_1 (s character) orbital. Since the p–s energy gap is considerable, the correlation is drawn rather steeply. The considerations of Rule 2 are not relevant.

Triatomic Molecules

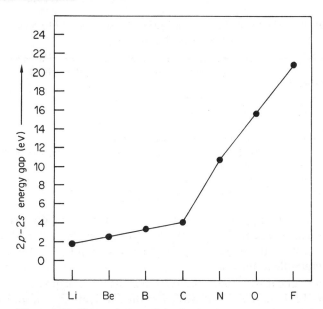

Figure 8.18. The variation of the 2p–2s energy gap along the first short period

In Figure 8.17 the position of the a_1 (s) orbital is by no means certain. The p–s gap varies with the nuclear charge of the atom, and those for the first short period are plotted in Figure 8.18.

Towards the beginning of the series the p–s gap is not very large and it is possible for the a_1 (s) orbital to be in the region of the a_1 and b_2 bonding orbitals. As the series is traversed the p–s gap becomes much larger and it is definite that the a_1 (s) orbital will occur at very much lower energies than the bonding levels. This point must be remembered when discussing the application of the Walsh Diagram (Figure 8.17) to molecules in which the charge on the nucleus of the central atom varies. As the charge increases the slope of the $\pi_u - a_1$ (s) correlation increases, and this has a significant effect in determining the shape of the AH_2 molecule.

(d) *Application of the Walsh Diagram to AH_2 Molecules*

The implications of the Walsh Diagram (Figure 8.17) as far as the shape of a molecule, AH_2, is concerned depend upon whether the various orbitals are more stable in the linear or bent state. A summary of this factor is given in Table 8.9.

Abiding by the principle of allowing the electrons of a molecule to occupy the lowest available energy levels, we can proceed to discuss the shapes of the molecules of Table 8.1 in terms of the Walsh Diagram.

Table 8.9. The Shape Factors Favouring the Stability of Molecular Orbitals in AH_2 Molecules

Orbital	Shape factor giving the more stable orbital (i.e. lowest energy)
$\overline{\sigma_u - b_2}$	Linear molecule
$\overline{\sigma_g - a_1}$	Linear molecule
$\pi_u - b_1$	—(Energy independent of shape)
$\pi_u - a_1\,(s)$	Bent molecule (strongly dependent upon bond angle)
$\sigma_u - b_2$	Linear molecule
$\sigma_g - a_1$	Linear molecule

Molecules with up to four electrons to distribute in the orbitals of Table 8.9 will be linear. Both of the lowest orbitals ($\sigma_g - a_1$ and $\sigma_u - b_2$) are most stable in the linear state. The linearity of BeH_2 (and $Hg(CH_3)_2$) is understandable in these terms.

With the radical BH_2 there are five electrons to be disposed of in addition to the $1s^2$ lone pair. Four will enter the low lying $\sigma_g - a_1$ and $\sigma_u - b_2$ orbitals, the fifth occupying the $\pi_u - a_1\,(s)$ orbital. This latter is very much stabilized by bending and therefore the molecule is expected to be bent. The bending will be slight because the other four electrons in the configuration,

$$(\sigma_g - a_1)^2 (\sigma_u - b_2)^2$$

are stabilized in a linear molecule. It is due to the slope of the $\pi_u - a_1\,(s)$ correlation that the bending effect of one electron can overcome the tendency towards linearity of the other four electrons. It is not possible to say how bent the molecule should be since Figure 8.17 is not at the moment quantitative.

The singlet state of the CH_2 molecule would be expected to have the configuration,

$$(\sigma_g - a_1)^2 (\sigma_u - b_2)^2 (\pi_u - a_1)^2$$

and with two electrons in the $\pi_u - a_1$ orbital is predicted to be more bent than BH_2. This is true. There is a triplet state of CH_2 which would have the configuration,

$$(\sigma_g - a_1)^2 (\sigma_u - b_2)^2 (\pi_u - a_1)^1 (\pi_u - b_1)^1$$

in which one electron occupies the strongly shape-dependent orbital, the other singly occupied orbital being the shape-independent $\pi_u - b_1$ non-bonding combination. It would, therefore, be expected to have a larger

Triatomic Molecules

bond angle than the singlet state and is, in fact, linear. Here the single electron in the $\pi_u - a_1$ orbital is not able to overcome the linear tendency of the lower four electrons. A contributing factor is undoubtedly the electron correlation between the two unpaired electrons. Thus stabilization will be a maximum when the triplet state is linear.

In the NH_2 molecule there are seven electrons to be accommodated in the 'valency' orbitals. These will have the configuration,

$$(\sigma_g - a_1)^2 (\sigma_u - b_2)^2 (\pi_u - a_1)^2 (\pi_u - b_1)^1$$

The molecule possesses two electrons in the critical $\pi_u - a_1$ orbital and would be expected to be bent to about the same extent as CH_2 (singlet), as is the case. The odd $\pi_u - b_1$ electron has no effect upon the shape of the molecule.

In the water molecule the eight electrons would have the configuration,

$$(\sigma_g - a_1)^2 (\sigma_u - b_2)^2 (\pi_u - a_1)^2 (\pi_u - b_1)^2$$

and again no significant bond angle change (from that in NH_2) would be expected. Again the prediction is in accord with fact.

So, it is seen that the Walsh approach to the bonding and shapes of these simple AH_2 molecules is highly successful.

The Sidgewick–Powell approach has some success but cannot explain all the cases satisfactorily.

The Walsh theory, which involves consideration of the antibonding orbitals of the molecules can make qualitatively successful predictions of the absorption spectra of these AH_2 species.

As examples of spectroscopic predictions consider the data in Table 8.10 for the regions in which three of the molecules considered absorb radiation. The data in Table 8.10 represent the lowest energy transitions of the molecules and these may be linked with electronic configuration and bond angle changes.

The BH_2 molecule has the ground state,

$$(\sigma_g - a_1)^2 (\sigma_u - b_2)^2 (\pi_u - a_1)^1$$

Table 8.10. Regions of Absorption by AH_2 Molecules

Molecule	Absorption region
BH_2	600–900 nm
NH_2	450–740 nm
OH_2	150–200 nm

and this can undergo an electronic transition in which the unpaired electron is excited to the $\pi_u - b_1$ non-bonding orbital. The configuration of the excited state is then:

$$(\sigma_g - a_1)^2 (\sigma_u - b_2)^2 (\pi_u - b_1)^1$$

which leaves the bonding electrons intact and should be a linear molecule with the bonds having the same length as in the ground state. These predictions are in accord with fact. The first excited state of BH_2 *is* linear and has the same bond length as in the ground state. The absorption producing the excited state is a measure of the

$$(\pi_u - b_1) - (\pi_u - a_1)$$

energy gap in BH_2. This is reasonably small as can be seen from Figure 8.17 for a bond angle of 131°.

The NH_2 molecule has the configuration,

$$(\sigma_g - a_1)^2 (\sigma_u - b_2)^2 (\pi_u - a_1)^2 (\pi_u - b_1)^1$$

in its ground state. The bond angle is less than that in BH_2 and the corresponding transition would be expected to occur at higher energies (shorter wavelengths), as is the case. The excited state configuration is

$$(\sigma_g - a_1)^2 (\sigma_u - b_2)^2 (\pi_u - a_1)^1 (\pi_u - b_1)^2$$

The bond angle in the excited state is 144° which is greater than that in the ground state (103·2°) and is in accordance with expectation.

The water molecule has a filled $(\pi_u - b_1)$ orbital and the corresponding transition cannot occur. The lowest energy transition would be from the higher non-bonding orbital $(\pi_u - b_1)$ to the lower of the two antibonding orbitals $(\overline{\sigma_g} - \overline{a_1})$. It would be expected to be of much higher energy than in BH_2 or NH_2, and corresponds to the configuration,

$$(\sigma_g - a_1)^2 (\sigma_u - b_2)^2 (\pi_u - a_1)^2 (\pi_u - b_1)^1 (\overline{\sigma_g} - \overline{a_1})^1$$

in which the bond angle should be larger than in the ground state $[(\overline{\sigma_g} - \overline{a_1})$ is stabilized in the linear state] and the bond length should be greater (because of the presence of an antibonding electron). The predictions are borne out by observation.

It is possible, with the development of photoelectron spectroscopy (Section 2.6), that the Walsh Diagram can be made more quantitative. For instance, the molecular orbitals of the water molecule are so arranged as to indicate four ionization potentials (apart from the $1s^2$ level). Three of these have been measured and found to be 12·61, 14·23 and 18·07 ev. They correspond to the energies required to remove electrons respectively

Triatomic Molecules

from the $(\pi_u - b_1), (\sigma_g - a_1)$ and $(\sigma_u - b_2)$ orbitals. The $(\pi_u - a_1)$ orbital, presumably, is at lower energy and the corresponding ionization potential must be greater than 21·21 ev, which is the limit of measurement at the present time.

The bond angle in the water molecule is 104·5° and therefore the bonding is intermediate between the sp (180°) and p–p (90°) extremes of the Walsh treatment. It should be pointed out that a more sophisticated treatment using two equivalent orbitals suitably constructed from s and p orbitals of the oxygen atom and the $1s$ orbitals of the hydrogen atoms has been carried out. It is consistent with the observed bond angle and with the observations of photoelectron spectroscopy.

8.3 The Sidgewick–Powell Approach to AB_2 Molecules

In the consideration of AB_2 molecules where B is an atom containing s and p orbitals (unlike the AH_2 case where the terminal hydrogen atoms possessed only $1s$ orbitals which could be used in bonding), it is possible for the p orbitals to be used in sigma and pi bonding to the central atom A. Some examples of such molecules will be dealt with, using first the Sidgewick–Powell treatment and then the Walsh treatment (Section 8.4).

The prediction of the shape of an AB_2 species depends upon the number of electron pairs surrounding the central atom which are occupying σ-type orbitals. Some examples in which pi bonding orbitals are occupied will be dealt with later in this section. For those molecules which contain only sigma electrons, the Sidgewick–Powell scheme is that they will take up positions of minimum repulsion. The results of such considerations are summarized in Table 8.11. Since, in AB_2 molecules, only two of the electron pairs need to be bonding, the remainder, $n-2$, occupy non-bonding orbitals. Table 8.12 summarizes the various examples chosen for the Sidgewick–Powell treatment.

Table 8.11. Spatial Distributions of $n(\sigma)$ Electron Pairs around a Central Atom (A)

n	Shape
2	Linear
3	Trigonal plane
4	Tetrahedral
5	Trigonal bipyramid
6	Square bipyramid (or octahedral)

Table 8.12. The Effect of Lone Pairs upon the Shapes of some Triatomic Molecules

Total electron pairs (n)	Number of lone pairs ($n-2$)	Predicted shape	Example
2	0	Linear	$HgCl_{2(g)}$
3	1	Bent, 120°	$SnCl_{2(g)}$
4	2	Bent, 109° 28′	$TeBr_{2(g)}$
5	3	Linear	ICl_2^-

The mercury atom has the ground state configuration, $6s^2\ 5d^{10}$, and becomes divalent after an s to p promotion. The two bonding pairs (sp hybrids) in mercuric chloride repel each other to a position of minimum repulsion in which the bond angle is 180°. The tin atom has the ground state configuration, $5s^2\ 5p^2$, and as such is divalent. The two unpaired electrons (in sp^2 hybrid orbitals) and the lone pair produce a trigonally planar distribution with a bond angle predicted to be just less than 120°.

In tellurium bromide there are two bonding pairs and two lone pairs thus producing a tetrahedral distribution and a bond angle just less than 109° 28′.

In the dichloroiodide ion, ICl_2^-, the iodine atom is surrounded by ten electrons. The iodine atom possesses seven valency electrons and these plus one from each of the chlorine atoms and plus the extra electron which makes the molecule negatively charged make ten. These electrons, arranged in electron pairs, will distribute themselves in a trigonal bipyramidal arrangement with the three lone pairs in the trigonal plane and the bonding pairs at an angle of 180° to each other thus producing a linear molecule. It is not necessary to bring in hybridization in any Sidgewick–Powell treatment as it is not essential to the arguments used. It is possible, of course, to rationalize the atomic orbital arrangement of the central atom in terms of hybridization to fit the predicted shape. With ICl_2^- the hybridization scheme would involve sp^3d hybrid orbitals. The ground state of iodine is $(Kr)\ 5s^2\ 4d^{10}\ 5p^5$ and that of I^- is the closed shell (Xe) structure, $5s^2\ 4d^{10}\ 5p^6$. In order that I^- should be divalent, a p to d promotion is invoked to give the configuration,

$$(Kr)\ 5s^2\ 4d^{10}\ 5p^5\ 5d_{z^2}^1$$

and this valency state leads to sp^3d hybridization as a means of producing five equivalent orbitals, two of which are used in bonding.

Triatomic Molecules

Examples of AB_2 Molecules Involving the Use of Pi Bonds

The Sidgewick–Powell treatment of AB_2 molecules which involve the use of pi bonds is based upon the spatial distribution of the sigma electron pairs. This means that those pairs of electrons which exist in pi bonding orbitals are discounted in determining the number of repelling pairs of electrons. Alternatively, since the pi bonding electrons exist in the same region within the molecule as do a sigma bonding pair, the four electrons forming the σ, π double bond between two atoms may be considered as one unit as far as repulsion is concerned.

The arguments involved will be exemplified in a discussion of the structures and shapes of the species, CO_2, NO_2^+, N_2O, N_3^- and UO_2^{2+}, which experimentally are found to be linear, and NO_2, NO_2^-, SO_2, O_3 and ONCl, which are bent. A summary of possible sigma electron pair distributions is shown in Table 8.13.

Table 8.13. Some Sigma Electron Pair Distributions in Molecules Containing Sigma and Pi Bonds

Examples	Number of electron pairs surrounding central atom				Spatial distribution of sigma pairs
	Total	Lone	Sigma	Pi	
HCN	4	0	2	2	Linear
ONCl, O_3	4	1	2	1	Trigonal plane
CO_2	4	0	2	2	Linear
SO_2	5	1	2	2	Trigonal plane

In CO_2 the two localized bonds may be regarded as being made up from σ, π pairs and, in consequence, the molecule is linear as are the isoelectronic NO_2^+, N_3^- and N_2O species. A more detailed consideration of the bonding in these linear molecules is to be found in Section 8.5.

The uranyl ion, UO_2^{2+}, is linear since in its +6 oxidation state the uranium atom has the valency configuration $7s^2$ which upon excitation to the divalent state becomes $7s^1 7p^1$ leading to the presence of two sigma bonding pairs of electrons. (An alternative promotion which is perhaps of lower energy is to one of the $5f$ orbitals—this does not affect the prediction of the shape of the ion.)

The molecule of nitrosyl chloride, ONCl, is bent and may be formulated with one double and one single bond as in Figure 8.19, there being a lone pair of electrons in one of the orbitals of the nitrogen atom. The valency shell of the nitrogen atom contains four electron pairs, two sigma

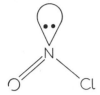

Figure 8.19. Electron pair distribution in the nitrosyl chloride molecule

bonding, one pi bonding and one lone pair, the repulsion to minimum positions between which results in a structure based upon a trigonal plane.

Other bent species such as O_3 and NO_2^- are isoelectronic with ONCl and their shapes are similar. The SO_2 molecule may be formulated as involving four-valent sulphur, the valency state of the sulphur being

$$(Ne)\ 3s^2\ 3p_x^1\ 3p_y^1\ 3p_z^1\ 3d_{z^2}^1$$

which may be utilized to form two sigma (sp^2) bonds and two pi bonds (pd hybrids), leaving a lone pair. There are five electron pairs to consider with the σ, π pairs existing in two bonding regions with the lone pair repelling both of these. The resulting structure is one based upon a trigonal plane causing the molecule to be bent.

It is instructive to consider the closely linked species NO_2^+, NO_2 and NO_2^- in terms of the Sidgewick–Powell treatment. The nitrite ion may be written formally as in Figure 8.20, and, with three repelling regions,

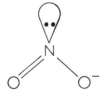

Figure 8.20. A formal structure for the nitrite ion, NO_2^-

should have a bond angle in the region of 120°. The removal of an electron reduces the repelling power of the non-bonding orbital and the bond angle would be expected to increase considerably in the radical NO_2. The removal of the remaining non-bonding electron produces the linear NO_2^+ ion.

The experimental values for the bond angles are shown in Table 8.14.

Without going into the electronic configuration in any great detail, the Sidgewick–Powell treatment predicts satisfactorily the dependence of the bond angle in terms of whether the lone pair orbital is vacant (180°), singly occupied (143°) or doubly occupied (115°).

Triatomic Molecules

Table 8.14. Bond Angles of some NO_2 Species

Species	ONO
NO_2^+	180°
NO_2	143° ± 11°
NO_2^-	115°

A note is required at this stage concerning the writing down of structures such as the one above for the NO_2^- ion in which the two bonds, although both between N and O atoms, are indicated to have different bond orders.

In the case of the nitrite ion, one bond is a single bond to an oxygen atom with a negative charge and the other is a double bond to a neutral oxygen atom. It must be understood that this does not represent the true electronic configuration of the nitrite ion. In terms of valence bond theory (see Section 9.8) such a structure would be that of one of two canonical forms, the wave functions of which form the basis for the total wave function. The two canonical forms of the nitrite ion may be written as in Figure 8.21.

The total molecular wave function would be a linear combination of ψ_1 and ψ_2, i.e.

$$\psi_{molecular} = \frac{1}{\sqrt{2}}(\psi_1 + \psi_2) \qquad (8.28)$$

indicating a 50 per cent contribution from each form to the total. The consequences of this *resonance*, as it is called, are that the two bonds are identical with bond orders of 1·5 and that the negative charge is spread evenly over the two oxygen atoms. It does not involve the spreading of the negative charge over the whole molecule as would seem possible when it is realized that nitrogen is almost as electronegative as oxygen. The molecular orbital treatment of such species is carried out in the next section which deals with the Walsh approach to AB_2 molecules.

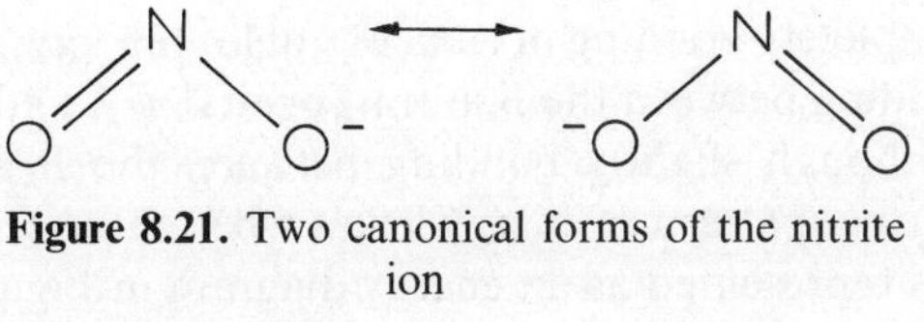

Figure 8.21. Two canonical forms of the nitrite ion

8.4 The Walsh Treatment of AB_2 Molecules

The principles involved in dealing with AB_2 molecules are the same as those used for AH_2 molecules. The differences arise because the B atoms possess p orbitals which may be used in sigma *and* pi bonding. The procedure is to set up the orbitals for linear and bent (90°) molecules as was done for AH_2 molecules in Section 8.2, and to correlate the two sets using the two Walsh Rules.

The Orbitals of Linear AB_2 Molecules

The linear molecules have orbitals which are sigma or pi in character and are either gerade or ungerade with respect to inversion. The s orbitals of the B atoms are taken to be non-bonding. They may be hybridized with the p orbitals under certain circumstances (low s–p energy gap) but since atom B is usually fairly electronegative this does not usually arise.

The two bonding orbitals are similar to those for the AH_2 case, except that they are constructed from p orbitals from the B atoms together with the s and one of the p orbitals of atom A, as shown in Figure 8.22. They

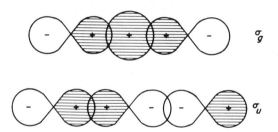

Figure 8.22. The two bonding orbitals of a linear AB_2 molecule

are both sigma orbitals but differ with respect to their inversion properties.

The other p orbitals of the three atoms are perpendicular to the molecular axis and may be arranged such that there are two sets directed along the x and y coordinates (the z coordinate taken to be the molecular axis). It is possible to overlap each of these sets to give three molecular orbitals. The diagrams of the overlaps pertaining to the three orbitals are shown in Figure 8.23.

There is a completely bonding orbital, ψ_1, of low energy, a non-bonding (slightly antibonding between the B atoms) orbital, ψ_2, and a high energy antibonding (although slightly bonding between the B atoms) orbital, ψ_3. This is the usual pattern of energy levels when three centre bonds are produced and is represented as an energy diagram in Figure 8.24.

Triatomic Molecules

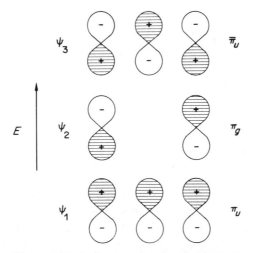

Figure 8.23. A diagram showing the method of overlapping of three p_x orbitals to give three pi molecule orbitals in a linear AB_2 molecule

The orbitals, ψ_1, ψ_2 and ψ_3, are all doubly degenerate (the two sets of p orbitals in the x and y directions produce two degenerate sets of pi orbitals), and are of pi symmetry with respect to the molecular axis. They can be further classified in terms of their inversion properties.

The ψ_1 and ψ_3 orbitals are ungerade while the non-bonding ψ_2 orbitals are gerade. Such characterization of the orbitals is also shown in Figures 8.23 and 8.24. Higher in energy than the pi orbitals are the antibonding sigma orbitals which may be drawn as in Figure 8.25, the $\overline{\sigma_g}$ orbital being

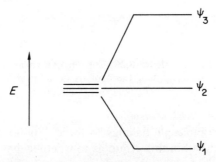

Figure 8.24. The relative energies of the orbitals ψ_1, ψ_2 and ψ_3 which are shown in Figure 8.23

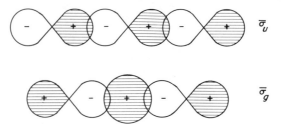

Figure 8.25. Overlap diagrams for the antibonding sigma orbitals of the linear AB_2 molecule

lower in energy than the $\bar{\sigma}_u$ orbital since it contains the s orbital character, whereas the latter is constructed from pure p orbitals.

As σ-type overlap is usually more efficient than π-type and as the overlap integral is closely connected with the bond strength, it may be said that of the sigma and pi bonding orbitals the former will be the lower in energy. The opposite will be the case for the antibonding orbitals, and the approximate order of energies for the orbitals of the linear AB_2 molecules is shown in Figure 8.26.

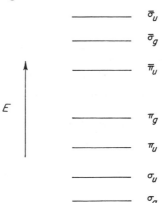

Figure 8.26. The relative energies of the orbitals of a linear AB_2 molecule (qualitative)

The Orbitals of 90° AB_2 Molecules

The bent (90°) molecule belongs to the C_{2v} point group. Its orbitals may be classified within this group as to whether they are symmetric (*a*) or antisymmetric (*b*) to the operation C_2 respectively, or to the operation σ_v (reflexion in the xz vertical plane) as subscript 1 or 2 respectively (see Table 8.6).

Triatomic Molecules

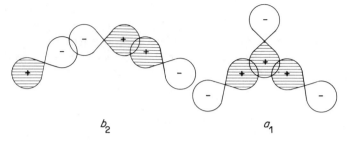

Figure 8.27. The two sigma bonding orbitals of a bent AB_2 molecule

The two σ-type bonding orbitals are shown in Figure 8.27.

A three centre system is possible for the p orbitals which are perpendicular to the molecular plane. Such orbitals are drawn in Figure 8.28.

The lowest of these, the bonding orbital, belongs to the b_1 representation and correlates with one of the lower π_u orbitals of the linear molecule. The non-bonding combination, a_2, correlates with one of the π_g orbitals of the linear molecule and the highest, antibonding orbital, $\overline{b_1}$, correlates

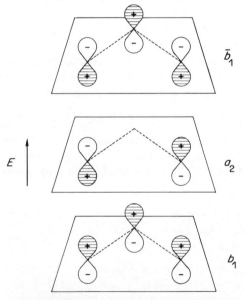

Figure 8.28. The three π orbitals of a bent AB_2 molecule. They are produced by the overlap of p_x orbitals which are perpendicular to the molecular plane

with one of the $\overline{\pi_g}$ orbitals. The degeneracy of the pi orbitals in the linear molecule is broken when the molecule bends, since the pi overlap between the p_y orbitals in the molecular plane is no longer feasible.

There will be an s orbital on atom A which is non-bonding in the bent state and which will correlate with the other π_g non-bonding orbital in the linear state. Other non-bonding orbitals in the bent state are the symmetric (a_1) and antisymmetric (b_2) combination of two p orbitals lying in the molecular plane as shown in Figure 8.29.

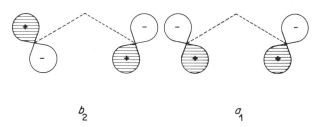

Figure 8.29. The symmetric and antisymmetric combinations of the p_y orbitals of the B atoms in a bent AB_2 molecule. Both represent non-bonding orbitals

The remaining orbitals are the two antibonding σ-type combinations as shown in Figure 8.30.

To complete a correlation of these orbitals with those of the linear state, it is best considered in terms of the effects of bending upon the orbitals of the linear state together with the correlations already mentioned.

The σ_g orbital correlates with an orbital (a_1) in the bent state which contains no s character. Rule 1 (see Section 8.2(c)) would indicate that the energy of the $\sigma_g - a_1$ orbital should rise with bending. Rule 2 indicates the opposite effect since in this orbital the B atoms are bonding, and a

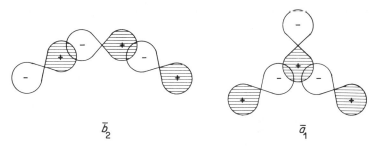

Figure 8.30. The two antibonding sigma orbitals of a bent AB_2 molecule

certain amount of stabilization is achieved by bending. This effect is the minor one and we may draw the energy of the $\sigma_g - a_1$ orbital as rising with decreasing bond angle.

The σ_u orbital correlates with the b_2 orbital. Both are of p character, but upon bending there will be an increase in energy since the orbital is antibonding with respect to the B atoms.

One of the π_u orbitals correlates with the b_1 orbital of the bent state (Figure 8.28) and both the π_u and b_1 combinations are entirely of p orbital character. Rule 2 indicates, however, that since the orbital is $B \leftrightarrow B$ bonding it should be stabilized upon bending.

The other π_u orbital (with p orbitals in the molecular plane) correlates with the a_1 non-bonding orbital (Figure 8.29) in the bent state. This orbital is also $B \leftrightarrow B$ bonding in the linear and bent states, but upon bending the $A \leftrightarrow B$ bonding character is lost and for this reason the $\pi_u - a_1$ orbital is shown to increase in energy slightly as the molecule bends.

The degenerate π_g non-bonding orbitals correlate with the a_2 (Figure 8.28) and b_2 (Figure 8.29) orbitals respectively. Both orbitals conserve their p character upon bending and both are $B \leftrightarrow B$ antibonding b_2 possibly to a greater extent than a_2 in the bent molecule. Both orbitals, therefore, are shown to increase in energy as the molecule bends.

The $\overline{\pi_u}$ antibonding doubly degenerate orbitals correlate with the non-bonding a_1 (s) orbital and the antibonding $\overline{b_1}$ (Figure 8.28) orbital. The $\overline{\pi_u}$ orbitals are of pure p character and so the $\overline{\pi_u} - a_1$ (s) orbital is shown to decrease rapidly in energy as the molecule bends (consequence of Rule 1). The $\overline{\pi_u} - b_1$ orbital does not vary in energy by very much upon bending, since the p orbital character is conserved. The orbital is, however, $B \leftrightarrow B$ bonding and in consequence is shown to stabilize slightly with decreasing bond angle.

The $\overline{\sigma_g}$ orbital correlates with the $\overline{a_1}$ (Figure 8.30) orbital. The $\overline{a_1}$ part contains no s character and therefore by Rule 1 there is an increase of energy in going from $\overline{\sigma_g}$ to $\overline{a_1}$ which is offset partially because the orbital is $B \leftrightarrow B$ bonding. The $\overline{\sigma_u} - \overline{b_2}$ orbital energy increases with bending since it is $B \leftrightarrow B$ antibonding.

The complete correlation diagram is shown in Figure 8.31. We can now use this diagram to discuss the electronic configurations and shapes of the AB_2 molecules already mentioned in Section 8.3 in which the Sidgwick–Powell treatment was applied.

Molecules containing sixteen or less electrons in the valency shells of the atoms concerned are expected to be linear since the majority of the lower eight orbitals are more stable in the linear state. The two s lone pairs on the B atoms are included in these eight and appear on the correlation

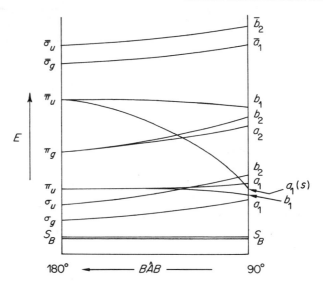

Figure 8.31. A Walsh diagram for AB_2 molecules. The orbital energies are plotted as a function of the bond angle. (After Walsh, *J. Chem. Soc.*, **1953**, 2268)

diagram in Figure 8.31. A molecule involving eighteen or twenty electrons would be expected to be bent since two electrons would occupy the $\overline{\pi_u}-a_1$ (s) orbital which is greatly stabilized by bending. The orbital energies would be minimized by a certain amount of bending. This bending would be reversed in the case of a twenty-two-electron molecule which would be expected to be linear since the additional electrons would occupy the $\overline{\sigma_g}-\overline{a_1}$ orbital. A summary of the shapes of molecules and ions and the number of their valency electrons is shown in Table 8.15.

Table 8.15. The Predicted and Observed Shapes of some AB_2 Molecules

Molecules	Number of valency electrons	Walsh prediction	Observed bond angle
CO_2	16	Linear	180°
NO_2^+, N_2O, N_3^-	16	Linear	180°
NO_2	17	Bent	143°
O_3	18	Bent	117°
SO_2	18	Bent	119°
NO_2^-	18	Bent	115°
OF_2	20	Bent	101°
I_3^-, ICl_2^-	22	Linear	180°

Triatomic Molecules

As can be seen from the examples in Table 8.15, the Walsh predictions are found to hold good for several molecules, ions and radicals. It now remains in this section for two examples to be dealt with in greater detail with reference to the distribution of electrons in the molecular orbitals. These examples are chosen to be the carbon dioxide molecule and the nitrite ion.

(a) *Carbon Dioxide*

The carbon dioxide molecule contains a total of twenty-two electrons, six of which occupy the virtually non-bonding $1s$ orbitals of the three atoms. The remaining sixteen may be distributed so as to minimize the energy of the molecule. The resulting electronic configuration is

$$S_B{}^2 S_B{}^2 (\sigma_g - a_1)^2 (\sigma_u - b_2)^2 (\pi_u - a_1)^4 (\pi_g - a_2)^2 (\pi_g - b_2)^2$$

and since the majority of these orbitals are stabilized in a linear state (as CO_2 is known to be) this configuration may be abbreviated to

$$S_B{}^2 S_B{}^2 \sigma_g{}^2 \sigma_u{}^2 \pi_u{}^4 \pi_g{}^4$$

where $S_B{}^2$ refers to a lone pair of $2s$ electrons on atom B ($=$ oxygen in this case). Of these orbitals the π_g are non-bonding, the rest being bonding. There are no antibonding electrons to be considered and, in consequence, there are four bonding pairs of electrons. If these are distributed so as to produce two identical bonds, the bond order of each bond will be two. This does not mean that the description of the bonding in terms of two localized double bonds is correct. The σ_g, σ_u, π_u and π_g orbitals are delocalized three-centre combinations. Experimental verification of the delocalized electronic configuration comes from the photoelectron spectrum of CO_2 which shows that there are four ionization potentials with values 19·29, 18·08, 17·23 and 13·68 ev corresponding to the ionizations of electrons from the σ_g, σ_u, π_u and π_g orbitals respectively.

The electron-in-a-potential-well approach also shows that, for the pi electrons, the delocalized arrangement is of lower energy than that involving two localized double bonds.

The energies of electrons in a one-dimensional potential well are given by equation (4.31). This equation may be applied to the two localized pi bonds in the following manner. The length of a localized double bond is b and the lowest energy level is that for which the quantum number, n, has a value of unity:

$$E_1 = \frac{h^2}{8mb^2} \tag{8.29}$$

Since both bonds are identical the four pi electrons occupy orbitals with one-electron energies given by equation (8.29). The total pi electron energy (ignoring electron repulsion) is $4E_1$. This is equal to $4h^2/8mb^2$ and will be referred to henceforth as $E_{localized}(E_L)$.

In the delocalized pi bond situation we have a 'well' of length $2b$, and the lowest level will have an energy given by

$$E_1 = \frac{h^2}{8m4b^2} \tag{8.30}$$

This may contain only two of the four pi electrons, the other two having to occupy the next level for which $n = 2$, and whose energy is given by

$$E_2 = \frac{4h^2}{8m4b^2} \tag{8.31}$$

The total pi electron energy is thus $2E_1 + 2E_2$ where E_1 is given by equation (8.30) and E_2 by equation (8.31). This will be called $E_{delocalized}$ (E_D), and is given by

$$E_D = \frac{2h^2}{8m4b^2} + \frac{8h^2}{8m4b^2} = \frac{10h^2}{32mb^2}$$

The ratio E_L/E_D is equal to $\frac{4}{8} \times \frac{32}{10} = 1 \cdot 6$, indicating that E_D is smaller than E_L and that the pi electrons in the CO_2 molecule will be more stable in delocalized orbitals.

In addition to the above considerations it may be pointed out that in the delocalized case, where the orbitals are larger, charge correlation will be less than in the localized case.

(b) *The Nitrite Ion*

The nitrite ion, NO_2^-, contains twenty-four electrons, six of these existing in the 1s orbitals of the three atoms. The other eighteen may be distributed in the lowest available molecular orbitals such that the molecule possesses minimum energy. The electronic configuration would be

$$S_B{}^2 S_B{}^2 (\sigma_g - a_1)^2 (\sigma_u - b_2)^2 (\pi_u - b_1)^2 (\pi_u - a_1)^2$$
$$[\overline{\pi_u} - a_1(s)]^2 (\pi_g - a_2)^2 (\pi_g - b_2)^2$$

the double occupation of the $\overline{\pi_u} - a_1 (s)$ orbital being the reason for its non-linearity.

All the pi type orbitals are delocalized over the three atoms and are three centre combinations. In addition to the σ-type bonds, then, there are two electrons in the $\pi_u - b_1$ orbital which is bonding, the remainder of the electrons being in non-bonding orbitals. The bond order of the two equivalent bonds in the nitrite ion is thus 1·5.

Triatomic Molecules

The transition from NO_2^- to NO_2 and to NO_2^+ involves two separate ionization processes and the opening up of the bond angle from 117° to 143° and eventually to 180°. The main orbital concerned in the production of a bent state is the $\overline{\pi_u} - a_1(s)$ combination and, if ionization occurs from this orbital, such opening of the bond angle as is observed would be explained. However, ionization occurs usually from the highest energy orbital in any system, and this would mean that in the bent nitrite ion the angle would be that where the $\overline{\pi_u} - a_1(s)$ orbital lies above the $\pi_g - a_2$ and $\pi_g - b_2$ orbitals. Ionization from these latter two orbitals would result in greater bending of the system. It is possible that photoelectron spectroscopy of NO_2 will provide experimental evidence of this point.

8.5 The Structures of BAC and HAB Molecules

Molecules in which two non-identical atoms are bonded to a central atom may be written as BAC (with A being the central atom) and as HAB in the special case of one of the atoms being a hydrogen atom. In these cases the linear molecules belong to the $C_{\infty v}$ point group and the bent molecules to the C_s point group which possesses only one element of symmetry which is the molecular plane. This means that the symmetry symbols for the linear molecules are either σ or π, but, since there is no inversion centre in $C_{\infty v}$ molecules, g and u subscripts do not apply. Also the orbitals in the bent state are classified as to whether they are symmetric or antisymmetric with respect to reflexion in the molecular plane as a' and a'' respectively.

The limitations, however, do not alter the orbital diagrams which apply just as well to BAC molecules as they do to AB_2 systems. Some modification is required in the case of HAB species since the hydrogen atom normally contributes only one orbital to the molecular orbital scheme. A separate diagram is required in this case, but one which is closely related to that for AB_2 (Figure 8.31). The modified diagram is shown in Figure 8.32.

From the way in which the orbital energies change with angle for BAC molecules (Figure 8.31) it is possible to predict that those containing sixteen valency electrons (N_2O, OCS, $ClCN$ and NCO^-) should be linear. The addition of more electrons brings into use the $\bar{\pi} - a'(s)$ orbital which is considerably stabilized in the bent molecules. Thus we would expect $ONCl$ and $ONBr$ to be bent in their ground states. They are.

We will mention only two HAB cases, namely hydrogen cyanide, HCN, and the hydroperoxy radical, HO_2. The latter is produced in the photolysis of hydrogen peroxide in aqueous solution, where the following reactions occur:

$$H_2O_2 \xrightarrow[253.7 \text{ nm}]{hv} 2OH$$

$$OH + H_2O_2 \longrightarrow H_2O + HO_2$$

$$2HO_2 \longrightarrow H_2O_2 + O_2$$

Hydrogen cyanide is a ten-electron molecule (ignoring the $1s^2$ pairs of the carbon and nitrogen atoms) and, as is found to be the case, should be linear with a bond of order three between the carbon and nitrogen atoms.

The hydroperoxy radical is a thirteen-electron molecule and would be expected therefore to be bent. Two electrons would occupy the $\bar{\pi} - a'(s)$ orbital which would produce the bending effect.

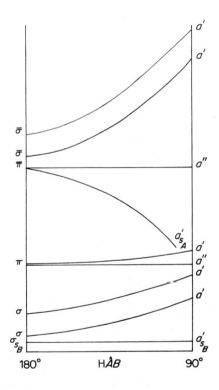

Figure 8.32. A Walsh diagram for HAB molecules. The orbital energies are plotted against the bond angle. (After Walsh, *J. Chem. Soc.*, **1953**, 2290)

Triatomic Molecules

8.6 The Molecules, SH_2 and PH_2

The bond angle in the hydrogen sulphide molecule is 92°. This is considerably less than that in water (104·5°). Similarly the ground state bond angle in the PH_2 radical (91°) is lower than that in the NH_2 radical (103·3°).

In terms of Walsh theory it would appear that the reductions in angle in going from the first short period element to the second short period element are due to a greater $\pi_u - a_1(s)$ slope relative to the slopes of the $\sigma_u - b_2$ and $\sigma_g - a_1$ correlations.

The Sidgewick–Powell approach is to consider that the lower electronegativities of the second row elements allows less polarization of the A—H bonds, and consequently the bonding pair–bonding pair repulsion is reduced. However, it seems that changing the group of the element, A, (and therefore its electronegativity) does not alter the bond angle in AH_2 appreciably. One drawback of the Sidgewick–Powell approach to almost 90° molecules is that it can lead to the impression that sp^3 hybridization is involved. This is definitely not so for molecules which do not have tetrahedral angles.

8.7 Summary

The two approaches to the shapes of triatomic molecules, the Sidgewick–Powell and the Walsh, have been treated in great detail. The latter approach has been seen to yield the maximum amount of information about these molecules and has considerable predictive use. The Sidgewick–Powell approach is less sophisticated and does not give very good predictions in some cases. However, its concepts are very simple compared to those of the Walsh treatment, and it is more generally applicable to polyatomic molecules.

It is hoped that the status of hybridization has been made obvious. The Sidgewick–Powell approach does not need the concept in order to predict molecular shapes. It may be added to the theory *after* the shape has been successfully predicted so that the orientations of the original atomic orbitals become reconciled with the shape.

In Walsh theory hybridization of atomic orbitals is not necessary and is only of use in the description of localized orbitals.

To predict the shape and bonding of a molecular system from hybridization theory is a misleading and incorrect procedure.

Problems

8.1 Discuss the shapes and bonding of the following molecules, applying the principles outlined in this chapter wherever possible:
  (a) N_2O
  (b) $HgCl_{2(g)}$
  (c) N_3^-
  (d) $AuCl_2^-$
  (e) O_3
  (f) ClO_2
  (g) F_2O
  (h) $IClBr^-$
  (i) the first excited state of CO_2
  (j) the first excited state of O_3
  (k) CS_2
  (l) C_3
  (m) CN_2^{2-}
  (n) S_3^{2-}
  (o) H_2Se
  (p) $HOCl$

8.2 Consider the description of the nitrosyl chloride molecule given in this chapter and discuss the possibility that the nitrogen–chlorine bond could have a bond order greater than one.

9

Other Polyatomic Molecular Systems

This chapter is devoted to the explanations of the bonding and shapes of more complex polyatomic molecular systems. In the majority of cases the Sidgewick–Powell approach is used for purposes of shape prediction. The molecular orbitals used in the various systems are built from the appropriate atomic orbitals and wherever possible the relevance of the Walsh approach is indicated.

9.1 Sidgewick–Powell Treatment of AB_n Molecules where $n \geqslant 3$

The principles of this treatment are that the sigma electron pairs in the valency shell of the central atom (A) arrange themselves in positions of minimum repulsion and that the order of the different types of repulsion is:

$$l \leftrightarrow l > l \leftrightarrow b > b \leftrightarrow b$$

l representing a lone pair and b representing a bonding pair of electrons.

The shapes of the AB_n molecules depend on the total number of sigma electron pairs and *not* just on the value of n (the number of bonding pairs). The basic arrangements of pairs of electrons around a central point are summarized in Table 9.1. Diagrams showing the minimum repulsion spatial arrangements of different numbers of electron pairs are shown in Figure 9.1.

Table 9.1. Arrangements of Minimum Repulsion of n Electron Pairs (σ) around a Central Point

n	Spatial arrangement
3	Trigonally planar
4	Tetrahedral
5	Trigonally bipyramidal
6	Octahedral

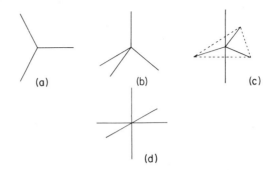

Figure 9.1. The spatial arrangements of n electron pairs ($n = 3, 4, 5$ and 6) surrounding a central atom corresponding to positions of minimum repulsion

As examples of compounds possessing the shapes mentioned in Table 9.1 we may deal with BF_3, CH_4, PCl_5, SF_6 and related compounds.

Boron Trifluoride, BF_3

In boron trifluoride there are six 'valency' electrons surrounding the boron atom and, in the form of three pairs, these will produce a trigonally planar arrangement of the atoms in the molecule with 120° bond angles (Figure 9.1(a)). This is the actual shape of the BF_3 molecule, although the bonding is more complex than that involving simple sp^2 hybridization. The reason for the latter statement is that the B—F bond lengths in the BF_3 molecule are 0·13 nm, whereas in the borofluoride ion, BF_4^-, the B—F bond lengths are considerably longer (0·153 nm). The BF_4^- ion is isostructural with methane, CH_4. These tetrahedral species have bonding which may be described in terms of sp^3 hybridization of the atomic orbitals of the respective central atoms and the consequent use of these to form four identical single σ-type bonds. The experimental fact that the B—F distance in BF_3 is lower than that in BF_4^- is an indication that the bond order of the bonds in BF_3 is greater than in BF_4^-. In other words there is evidence for multiple bonding in BF_3. If sp^2 hybridization is invoked on the boron atom there is a pure p orbital left which is perpendicular to the molecular plane. This may participate in the formation of a four-centre system of π-type molecular orbitals by combining suitably with one p orbital (codirectional) of each of the fluorine atoms. The four molecular orbitals produced are shown in Figure 9.2.

There is one bonding orbital, ψ_B, two degenerate non-bonding orbitals, ψ_N, and one antibonding orbital, ψ_{AB}. The pi electronic configuration in BF_3 would be

Other Polyatomic Molecular Systems

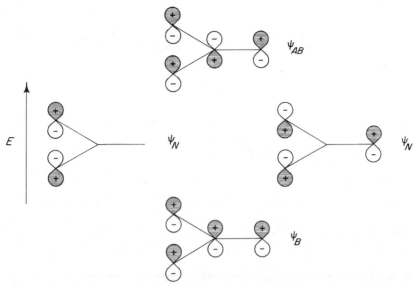

Figure 9.2. The four pi molecular orbitals produced from four p atomic orbitals in a trigonally planar AB_3 molecule

$$\psi_B{}^2 \psi_N{}^4$$

and there would be a B—F bond order of 1·33 since only one π electron pair is bonding and the bonding effect is spread over the three bonds. The introduction of this delocalized pi bonding does not alter the shape of the molecule as derived by applying the Sidgewick–Powell principles.

It is interesting to note that the delocalized π-type bonding suggested for BF_3 is not possible for the BH_3 molecule since hydrogen does not possess any p electrons. The BH_3 molecule has only a transient existence and the simplest hydride of boron has the formula B_2H_6, the structure and bonding of which is described below (Section 9.6).

The BF_3 molecule possesses six electrons in the valency shell of the boron atom and twenty-four valency electrons altogether. (The boron atom contributes three electrons and the fluorine atoms each contribute seven electrons.) Other twenty-four-electron molecules we can treat are the carbonate ion, nitrate ion and gaseous sulphur trioxide.

(a) *Carbonate Ion*, $CO_3{}^{2-}$

The tetravalent state of the carbon atom is $(He) 2s^1 2p_x{}^1 2p_y{}^1 2p_z{}^1$ which is achieved by an s to p promotion from the ground state. The

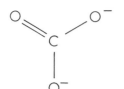

Figure 9.3. A representation of the bonding in the carbonate ion

carbonate ion may be written as having the structure shown in Figure 9.3, involving one double bond to oxygen and two single bonds to the oxygens carrying the negative charges. As we shall see later, this structure is not absolutely correct, but serves as a basis for the correct prediction of the shape of the ion.

As written, the ion contains three sigma bonds and one pi bond. For the sigma bonding we therefore have three electron pairs which will lead to a trigonally planar ion. The hybridization of the carbon orbitals will be sp^2 and the remaining p orbital will be used to form the pi bond. This latter does not alter the shape.

As far as the molecular orbital description is concerned the twenty-four electrons would be placed in molecular orbitals and, in particular, there would be six electrons in the sigma bonding orbitals and twelve electrons in the non-bonding orbitals on the *ligand* atoms (the B atoms in AB_3 are the ligand atoms), and the remaining six electrons occupy the delocalized four-centre molecular orbital system to give $\psi_B{}^2\psi_N{}^4$ (as with BF_3). The C—O bond order is thus 1·33 and the negative charges are also delocalized over the whole ion, there being a greater concentration on the oxygen atoms because of their greater electronegativity.

(b) *Nitrate Ion*, $NO_3{}^-$

The valency state of the nitrogen atom is (He) $2s^2\ 2p_x{}^1\ 2p_y{}^1\ 2p_z{}^1$, and as such possesses only three singly occupied orbitals. The nitrogen atom may become tetravalent by using these orbitals and additionally by using the remaining electron pair in the formation of a coordinate bond. A co-ordinate bond is an electron pair bond where the electron pair is supplied by one or other of the atoms participating in the bond. The structure

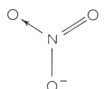

Figure 9.4. A representation of the bonding in the nitrate ion

of the nitrate ion may be represented as in Figure 9.4, which involves a single bond (σ) to the negative oxygen atom, a *coordinate* bond (σ) to one of the neutral oxygen atoms (in which the nitrogen atom supplies both electrons involved in the bonding) and a double σ, π bond to the third oxygen atom. With one of the *p* orbitals reserved for the pi bond the situation is one involving three sigma electron pairs surrounding the nitrogen atom which leads to the trigonal planarity of the nitrate ion. As with BF_3 and CO_3^{2-} the hybridization required is sp^2.

The molecular orbital approach would, as with the other examples, give a pi electron arrangement, $\psi_B^2 \psi_N^4$, with an N—O bond order of 1·33, and does not require the postulation of any coordinate bond.

(c) *Sulphur Trioxide Gas*

Gaseous sulphur trioxide consists of discrete SO_3 units, whereas the solid is more complex and contains trimeric molecules, $(SO_3)_3$. The ground state of the sulphur atom is $(Ne) 3s^2\ 3p_x^2\ 3p_y^1\ 3p_z^1$, and as such the sulphur atom would be divalent as it is in H_2S. The difference between sulphur and carbon and nitrogen is that it possesses low lying *d* orbitals which may be used in bonding. The *d* orbitals in carbon and nitrogen (3*d*) are of too high an energy to be used. It is possible with the sulphur atom to carry out $p \rightarrow d$ and $s \rightarrow d$ promotions to give the hexavalent state,

$$(Ne)\ 3s^1\ 3p_x^1\ 3p_y^1\ 3p_z^1\ 3d_{x^2-y^2}^1\ 3d_{z^2}^1$$

With the two *d* orbitals and one of the *p* orbitals reserved for the pi bonding, we are left with sp^2 type sigma bonding resulting in a trigonally planar molecule which may be written as shown in Figure 9.5. The pi bonding involves *p* orbitals from the three oxygen atoms, together with a *p* orbital and two *d* orbitals from the sulphur atom.

The molecular orbital treatment as applied to BF_3, CO_3^{2-} and NO_3^- would, of course, predict S—O bond order of 1·33 since the *d* orbital participation was not relevant in the former cases. If the *d* orbitals are taken into account there would be three pi bonding orbitals and three antibonding orbitals, the latter three being vacant in SO_3 and leading to a bond order of two which is consistent with observation. (The S—O bond length is 0·143 nm which is considerably shorter than the value

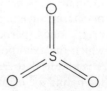

Figure 9.5. The localized bonding in the sulphur trioxide molecule

expected for a single bond of 0·170 nm.) The ψ_N^4 configuration of the BF_3, CO_3^{2-} and NO_3^- molecular systems is, in SO_3, replaced by the four electrons forming two of the three pi bonds, the four electrons existing in *bonding* orbitals.

Methane, CH_4

The carbon atom has a (He) $2s^2\, 2p_z^1\, 2p_y^1$ ground state which is divalent. Although the CH_2 molecule does exist transiently in coal gas–air flames, it is far less stable than the methane molecule, CH_4, in which the carbon is tetravalent. To attain this state an $s \rightarrow p$ promotion must occur to give the configuration, (He) $2s^1\, 2p_x^1\, 2p_y^1\, 2p_z^1$, which allows four electron pair bonds to be formed. The formation of such bonds involves the release of the bond energy which more than compensates for the use of the promotion energy.

The four electron pairs (all bonding and equivalent) would repel each other to the tetrahedral position of minimum repulsion, the bond angles being 109° 28'. The tetrahedral distribution of bonds around a carbon atom is, of course, of extreme importance in understanding the stereochemistry of many organic compounds. The sp^3 hybridization of the atomic orbitals of carbon is necessary to reconcile them with the tetrahedral shape of the CH_4 molecule.

Phosphorus Pentachloride, PCl_5

The phosphorus atom has the ground state configuration, (Ne) $3s^2\, 3p_x^1\, 3p_y^1\, 3p_z^1$, and as such is trivalent. It is the configuration used in the formation of the trichloride, PCl_3. In order that the phosphorus atom should be pentavalent, an $s \rightarrow d$ promotion is necessary to give the configuration,

$$\text{(Ne)}\ 3s^1\, 3p_x^1\, 3p_y^1\, 3p_z^1\, 3d_{z^2}^1$$

The five bonding pairs will distribute themselves in a trigonally bipyramidal manner, consistent with the minimization of electron pair repulsion. The hybridization of the atomic orbitals of the phosphorus atom required to make this arrangement feasible is sp^3d. This may be thought of as being a mixture of sp^2 hybridization to give the trigonal planar arrangement as usual and pd hybridization which supplies the axial bonds (perpendicular to the trigonal plane). The C_3 axis (along which the axial bonds are directed) may be taken to be the z coordinate axis and in that case the atomic orbitals used to form the pd hybrids are the p_z and d_z^2 orbitals which also are directed along the z axis.

Other Polyatomic Molecular Systems

Sulphur Hexafluoride

The hexavalent state of the sulphur atom has already been described (under SO_3). If all six valency electrons are used to form sigma bonds to the fluorine atoms in SF_6, there will be six bonding pairs surrounding the sulphur atom. They will arrange themselves in an octahedral manner in which they are directed towards the vertices of a regular octahedron so as to minimize the repulsions between electron pairs. The hybridization essential to explain this distribution is sp^3d^2, the d_z^2 and $d_{x^2-y^2}$ orbitals of the central atom being used.

9.2 The Effects of Non-Bonding Pairs

We will now consider the stereochemical consequences of including non-bonding pairs of electrons in the valency group around a central atom in AB_n molecules. The Sidgewick–Powell results are summarized in Table 9.2. The cases in which $m = 0$ have been dealt with in the previous section. We may now consider examples of those cases in which m has a non-zero value.

Table 9.2. Spatial Distributions of n Pairs of Sigma Bonding Electrons with m Pairs of Non-Bonding Electrons

n	m	$n+m$	Spatial arrangement
4	0	4	
3	1	4	Tetrahedral
2	2	4	
5	0	5	
4	1	5	Trigonally bipyramidal
3	2	5	
2	3	5	
6	0	6	
5	1	6	Octahedral
4	2	6	

$n = 3, m = 1$

The simplest example is the ammonia molecule, NH_3. The ground state of the nitrogen atom is also the trivalent state, $(He) 2s^2\ 2p_x^1\ 2p_y^1\ 2p_z^1$, and with sp^3 hybridization to explain the modification of the atomic orbitals to a tetrahedral distribution the lone pair will occupy one of the four positions leaving a trigonally pyramidal molecule. The bond angle in ammonia is, in fact, 107° and not 109° 28′, the explanation being that the lone pair–bond pair repulsion is greater than that between two bonding pairs.

The bond angle in NF_3 is reported to be 102° 9′ and that in PCl_3 to be 100° 6′, both of these molecules being examples of this class.

$n + m = 5$

Four of the n, m combinations in Table 9.2 lead to a total of five pairs of electrons and may be discussed together. All of these configurations lead to a trigonally bipyramidal spatial distribution of the five electron pairs. In some cases there are alternative arrangements possible for the bonding and non-bonding electron pairs. The $n = 5$ case is straightforward and has been dealt with (PCl_5).

For the $n = 4, m = 1$ case there are two possibilities, these being shown in Figure 9.6, which differ in the situation of the lone pair electrons. The lone pair may either be in an axial position (Figure 9.6(a)), or it may be

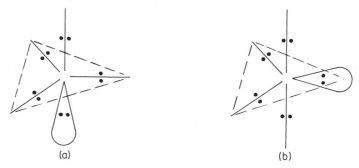

Figure 9.6. The two possible alternative locations of a lone pair of electrons in a distribution which is trigonally bipyramidal: (a) axial, (b) planar

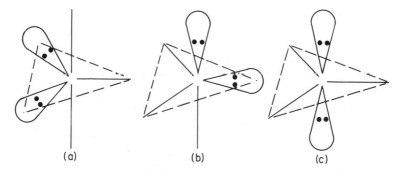

Figure 9.7. The three possible alternative arrangements of two lone pairs of electrons in a distribution which is trigonally bipyramidal: (a) both in the trigonal plane, (b) one axial and one in the trigonal plane, and (c) both in axial positions

Other Polyatomic Molecular Systems

in the trigonal plane (Figure 9.6(b)), and experimentally it is observed that the latter arrangement describes actual molecular structures in this class, the former arrangement presumably being of higher energy.

For the $n = 3$, $m = 2$ case, there are three possible arrangements of the bonding and lone pairs within the basic trigonal bipyramid, as is shown in Figure 9.7. By observation the most stable arrangement is found to be that in which both the lone pairs of electrons are located in the trigonal plane (Figure 9.7(a)).

There are three possible arrangements for the $n = 2$, $m = 3$ case which are shown in Figure 9.8. Again, as with the previous two cases, the most stable arrangement is the one in which the lone pairs are located in the trigonal plane (Figure 9.8(c)).

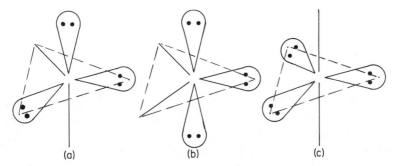

Figure 9.8. The three possible alternative arrangements of three lone pairs of electrons in a distribution which is trigonally bipyramidal: (a) two planar, one axial, (b) one planar, two axial, and (c) all three in the trigonal plane

As examples of molecules with these spatial arrangements of five electron pairs, we may take $TeCl_4$, ClF_3 and ICl_2^-.

(a) *Tellurium Tetrachloride*, $TeCl_4$

The ground state of the tellurium atom has the electronic configuration, $(Kr) 5s^2 5p_x^2 5p_y^1 5p_z^1$, and as such it is divalent. To make it tetravalent a $p \to d$ promotion is necessary to give the configuration,

$$(Kr) 5s^2 5p_x^1 5p_y^1 5p_z^1 5d_{z^2}^1$$

Four electron pair bonds may then be formed by using the unpaired electrons together with one electron from each of the four chlorine atoms, leaving a single lone pair in what may be described as sp^3d hybrid orbital.

The shape of the molecule would be based on a trigonal bipyramid with the lone pair in the trigonal plane as shown in Figure 9.9. The greater repelling effect of the lone pair causes the non-collinearity of the

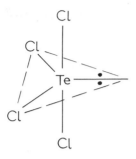

Figure 9.9. The location of the lone pair of electrons in the structure of the tellurium tetrachloride molecule

Cl—Te—Cl axial bonds, the shape being best described as a distorted tetrahedron.

(b) *Chlorine Trifluoride*, ClF_3

The ground state of the chlorine atom has the electronic configuration, $(Ne)\,3s^2\,3p^5$, which is monovalent, and with one $p \to d$ promotion the trivalent state,

$$(Ne)\,3s^2\,3p_x^2\,3p_y^1\,3p_z^1\,3d_{z^2}^1$$

is produced in which the chlorine atom can form bonds to the three fluorine atoms, there being also two lone pairs. The five pairs of electrons arrange themselves according to Figure 9.7(a), as is shown in Figure 9.10, with the lone pairs causing the FĈlF angle to be less than 90° (87°).

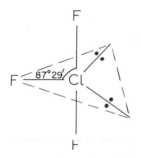

Figure 9.10. The location of the two lone pairs in the structure of the chlorine trifluoride molecule

(c) *Iodine Dichloride Ion*, ICl_2^-

The ground state of the iodine atom is monovalent, $(Kr)\,5s^2\,5p^5$, and if we place the negative charge (one electron) of the ion, ICl_2^-, on to the iodine atom for convenience, we can attain the divalent state if the extra electron occupies a d orbital giving, for the iodine atom, the electron configuration,

$$(Kr)\,5s^2\,5p_x^2\,5p_y^2\,5p_z^1\,5d_{z^2}^1$$

Other Polyatomic Molecular Systems

In the ion the two bonding pairs and three lone pairs arrange themselves as in Figure 9.8(c) and the ion is linear, there being no unsymmetrical lone pair effects.

The negative charge would not be localized on the iodine atom since chlorine is more electronegative.

$n + m = 6$

There are three n, m combinations in Table 9.2 which lead to a total of six pairs of electrons. The six pairs are spatially orientated towards the vertices of a regular octahedron, and unlike the trigonal bipyramidal arrangement all the directions are equivalent. No four bonds in a square plane are unique so there is no problem of placing *one* lone pair. It simply takes one of the octahedral positions, leaving the molecule with a square pyramidal shape, as shown in Figure 9.11.

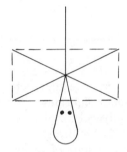

Figure 9.11. The location of a lone pair of electrons in a distribution which is octahedral

An example of a molecule with five bonding pairs of electrons and one lone pair is iodine pentafluoride, IF$_5$, where the five-valent state of the iodine atom is

$$(Kr)\ 5s^2\ 5p_x^1\ 5p_y^1\ 5p_z^1\ 5d_{x^2-y^2}^1\ 5d_{z^2}^1$$

The lone pair occupies an sp^3d^2 hybrid orbital (as do the bonding pairs), and is in one of the octahedral positions. The effect of the lone pair is to cause the central iodine atom to be slightly below the square plane containing the four fluorine atoms. In other words, the $\widehat{FIF}$ angle between the planar F—I bonds and the axial bond is slightly less than 90°.

If we now consider the case of six pairs of electrons including two lone pairs, we do have to consider the alternative positions of these lone pairs. The two possibilities are shown in Figure 9.12, and would predict (a) a square planar molecule, or (b) a distorted tetrahedron respectively. The more symmetrical square planar arrangement in which the lone pair–lone pair repulsion is minimized (Figure 9.12(a)) is found to be the more stable configuration, and an example is the ICl_4^- ion.

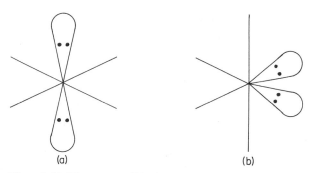

Figure 9.12. The two possible alternative arrangements of two lone pairs in a distribution which is octahedral: (a) *trans*, and (b) *cis*

9.3 Molecules with Multicentre Sigma Bonds

In many of the examples in the previous section, valency states have been postulated to involve low-lying d orbitals. An alternative approach which was developed after the preparation of the xenon fluorides is to use three centre bonding. The xenon difluoride molecule, XeF_2, is observed to be linear. It is also isoelectronic with ICl_2^- and the same bonding might be expected to be involved. The divalent state of xenon would be that with the electronic configuration,

$$(Kr)\ 5s^2\ 5p_x^2\ 5p_y^2\ 5p_z^1\ 5d_{z^2}^1$$

with one of the p electrons having been promoted to a d orbital. The five electron pairs in XeF_2 would be stable in a trigonally bipyramidal form with the three lone pairs in the trigonal plane resulting in the molecule being linear.

An alternative bonding scheme which does not involve the expenditure of promotion energy was suggested by Coulson. One p orbital from each of the three atoms in the molecule could be combined in a three-centre system to give three molecular orbitals which are drawn in Figure 9.13.

In XeF_2 these orbitals would be used to accommodate four electrons—two from the xenon atom and one from each of the fluorine atoms. The bond order of the Xe—F bonds would be $\frac{1}{2}$ and the bond energy therefore would be expected to be rather low. It is observed to be 30 kcal.mole^{-1} (125·5 kJ.mole^{-1}).

Other Polyatomic Molecular Systems

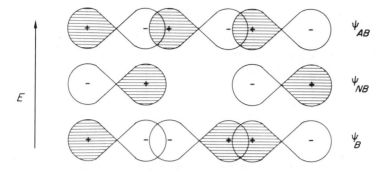

Figure 9.13. The three-centre molecular orbitals produced from three colinear p atomic orbitals

In the square planar XeF_4 molecule this kind of bonding scheme could be carried out for the p orbitals in, say, the x and z directions and, since p orbitals on any atom are mutually perpendicular, this would explain the shape of the molecule. The scheme can be carried one step further to predict that XeF_6 should be a regular octahedron. From an analysis of vibrational spectra it has been shown that the XeF_6 molecule is octahedral in the gas phase. The Sidgewick–Powell treatment would predict a structure which may be described as being a distorted octahedron. The xenon atom possesses eight valency electrons and these, together with the six electrons from the fluorine atoms, may form seven electron pairs. These would repel each other, the position of minimum repulsion being a pentagonal bipyramid. The lone pair would probably occupy a position in the pentagonal plane leaving the six bonding pairs arranged in a distorted octahedral manner. It should be pointed out, however, that the $TeCl_6^{2-}$ ion, which has the same number of valency electrons as XeF_6, is known to be regularly octahedral. There is evidence that the ions, $TeBr_6^{2-}$ and TeI_6^{2-}, are not regularly octahedral.

9.4 Molecules Involving Multiple Bonds

In the general case, we have to consider the shapes of molecules which not only have bonding and non-bonding pairs of electrons but also those in which some of the bonding pairs take part in pi bonding. As far as the Sidgewick–Powell treatment is concerned, the shape of a molecule is based on the positions of minimum repulsion of those electron pairs in σ-type orbitals. The pi electrons, to a first approximation, are ignored.

The commonest cases involve species of the AB_3 and AB_4 types.

AB₃ Molecules

In Section 9.1 the carbonate and nitrate ions were dealt with. It was seen that they possessed twenty-four valency electrons including eight electrons in the valency group of the central atom. Of these eight electrons, six were involved in the sigma bonding and two were delocalized in the bonding pi orbital. The doubly degenerate pi orbitals were also occupied and any increase in the total number of electrons in the molecule would, if the bonding system remained the same, involve the use of the high energy, ψ_{AB}, antibonding orbital. In preference to such an expenditure of energy, a change in the stereochemistry occurs. This conclusion may be drawn from a study of the Walsh Diagram for the appropriate system. In the case of triatomic molecules it is found that they are linear until electrons have to occupy an orbital which is strongly stabilized in the bent state. For an AB_2 molecule it is possible to accommodate up to sixteen electrons without using any orbitals which will cause any bending tendency. The seventeenth and eighteenth electrons must occupy the $\bar{\pi}_u - a_1(s)$ orbital which, in the linear molecule, is $\bar{\pi}_u$ and antibonding, but which is a low energy non-bonding a_1 orbital in the 90° molecule. As a result of this correlation, molecules with seventeen and eighteen electrons are bent.

The Walsh predictions for AB_3 systems are that those containing up to twenty-four electrons should be trigonally planar (cf. BF_3, CO_3^{2-}, NO_3^-) and those with twenty-five or twenty-six electrons should be trigonally pyramidal. For examples of these we may deal with the chlorine trioxide molecule, and the sulphite and iodate ions.

(a) *Chlorine Trioxide*, ClO_3

This may be described in terms of the seven-valent state of the chlorine atom, (Ne) $3s^1\ 3p^3\ 3d^3$, which is produced from the ground state by two $p \rightarrow d$ and one $s \rightarrow d$ promotions. The chlorine valency state may then use six of these electrons in forming σ, π double bonds to the three oxygen atoms. This would leave a singly occupied orbital in the fourth of the tetrahedral positions. The bond angle would be expected to be somewhat greater than 109° 28' and is observed to be 112°.

(b) *Sulphite Ion*, SO_3^{2-}

The bonding may be thought to be made up from two S—O⁻ bonds and an S=O double bond. The sulphur atom is thus in its four-valent state, (Ne) $3s^2\ 3p^3\ 3d^1$, and if the singly occupied d orbital is used in pi bond formation, the sigma hybridization about the central sulphur atom will be sp^3 and the ion will be pyramidal with the lone pair occupying one of the tetrahedral positions.

Other Polyatomic Molecular Systems

(c) *Iodate Ion*, IO_3^-

The iodine atom in its five-valent state has the configuration,

$$(Kr)\, 5s^2\, 5p_x^1\, 5p_y^1\, 5p_z^1\, 5d_{x^2-y^2}^1\, 5d_{z^2}^1$$

and it is possible for the two singly occupied d orbitals to be used in the pi bonding leaving sp^3 hybridization of the other orbitals to explain the tetrahedral distribution of the four electron pairs (one lone + three bonding) and the consequent pyramidal shape of the ion. The nature of the electronic configurations of such ions is separate from the Sidgewick–Powell description of the shapes of the ions.

The three bonds in the SO_3^{2-} and IO_3^- ions will have bond orders of $1\tfrac{1}{3}$ and $1\tfrac{2}{3}$ respectively. The bonds in each ion are identical and the negative charge is also equally spread as far as the oxygen atoms are concerned.

The Walsh description of AB_3 molecules indicates that those containing twenty-four valency electrons should be planar. The next lowest energy orbital (which is vacant in twenty-four-electron molecules) is the antibonding orbital already referred to, and which is shown in Figure 9.2. This orbital, ψ_{AB}, in the planar molecule correlates with a non-bonding s orbital on atom A in the pyramidal 90° molecule. Electrons in such an orbital become considerably stabilized when the molecule bends. Thus the Walsh treatment would predict that twenty-five- and twenty-six-electron molecules should be pyramidal with bond angles less than 120°. Examples of twenty-six-electron molecules are SO_3^{2-}, IO_3^-, PCl_3 and NF_3, in which the observed bond angles are in the region of 102–104°.

AB_4 Molecules

The methane molecule is tetrahedral and its structure may be explained in terms of four bonding pairs of electrons arranged so that electron pair repulsion is minimized.

Molecules involving multiple bonds which are also tetrahedral are the phosphate ion, PO_4^{3-}, sulphate ion, SO_4^{2-}, and perchlorate ion, ClO_4^-. All these ions are isoelectronic thirty-two-electron species and all are observed to be tetrahedral. The *ortho*-silicate ion, SiO_4^{4-}, which occurs naturally, is a thirty-two-electron ion which is regularly tetrahedral but which does not involve multiple bonding in a formal sense, in which all four bonds are electron pair bonds between the silicon atom and the negatively charged oxygen atoms. The valency state of the silicon atom is s^1p^3, and these orbitals may be sp^3 hybridized to fit with the observed shape. The Sidgewick–Powell treatment would predict a regular tetrahedral arrangement of the four electron pair bonds. As we shall see later, this does not necessarily give a good description of the molecule as some multiple bonding may occur.

In the *ortho*-phosphate ion, PO_4^{3-}, there is certainly some multiple bonding which may be represented by the arrangement shown in Figure 9.14. This is consistent with the five-valent state of phosphorus,

$$(Ne)\ 3s^1\ 3p_x^1\ 3p_y^1\ 3p_z^1\ 3d_{z^2}^1$$

in which four of the valency electrons are involved in sp^3 type sigma bonds and the fifth (the d electron) is involved in the pi bonding.

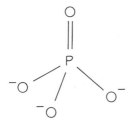

Figure 9.14. A representation of the bonding in the *ortho*-phosphate ion

In reality all the bonds are equivalent and the above treatment would predict a bond order of $1\frac{1}{4}$ for each P—O bond. In other words the pi electrons and the negative charges are delocalized.

In a like manner we may regard the sulphate ion as having the arrangement shown in Figure 9.15, involving two pi bonds in addition to the

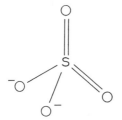

Figure 9.15. A representation of the bonding in the sulphate ion

tetrahedrally arranged sigma bonds. This is consistent with the six-valent state of sulphur in which the two d electrons are involved in the pi bonding and the sp^3 hybrid orbitals are used in the sigma bonding. The bond order of the S—O bonds would be $1\frac{1}{2}$ and the negative charge would be delocalized.

In the perchlorate ion, ClO_4^-, the seven-valent state of the chlorine atom is involved, there being three d orbitals used in the pi bonding in addition to the usual sp^3, tetrahedral sigma bonding. The bond order of the Cl—O bonds would be $1\frac{3}{4}$.

Other Polyatomic Molecular Systems

In SiO_4^{4-}, PO_4^{3-}, SO_4^{2-} and ClO_4^-, then, the ions are all *regularly tetrahedral* (position of minimum repulsion of the four sigma electron pairs), and the pi bonding has no effect on the stereochemistry since the pi electrons are delocalized over the whole molecule. This latter point will now be dealt with in more detail.

First there is experimental evidence to consider. The lengths of the bonds between the oxygen atoms and the central atoms in the ions considered above are shown in Table 9.3.

Table 9.3. A—O Bond Lengths in the Ions AO_4^{n-}

A	n	Bond length (nm)
Si	4	0·163
P	3	0·154
S	2	0·149
Cl	1	0·146

It has been estimated that the single bond (σ) distances in the above ions would be 0·176, 0·171, 0·169 and 0·168 nm respectively. The actual bonds are thus all shorter than expected for a single bond by 0·013, 0·017, 0·020 and 0·022 nm respectively. This indicates the operation of what may be called p_π—d_π bonding, the extent of it increasing from SiO_4^{4-} to ClO_4^- as expected from the foregoing simple treatment. The simple treatment did not predict p_π—d_π bonding in SiO_4^{4-} since no pi bonding was involved in the simple structure used to predict the shape, but nevertheless there is experimental evidence for its participation. The d_{z^2} orbital may overlap with four p orbitals which are themselves perpendicular to the A—O bond directions, thus forming a five-centre molecular orbital system.

The $d_{x^2-y^2}$ orbital may overlap with the other p orbitals on the oxygen atoms which are perpendicular to the A—O directions and to the p orbitals used in overlap with the d_{z^2} orbital.

In ClO_4^- the simple treatment based upon the seven-valent state of chlorine requires that, in addition to the d_{z^2} and $d_{x^2-y^2}$ orbitals, one of the d_{xy}, d_{xz} or d_{yz} orbitals should be used in the pi bonding.

(a) Trimethylamine and Trisilylamine

So far we have dealt with molecules involving multiple bonding where the pi bonding has not influenced the stereochemistry of the molecule. This is not always the case, however, and an example where the pi bonding has a definite influence upon the shape of a molecule will now be presented.

In trimethylamine, $N(CH_3)_3$, the shape of the molecule depends upon there being four electron pairs surrounding the nitrogen atom. Three of these are bonding pairs and the other is a non-bonding pair. We thus expect the molecule to be pyramidal, as shown in Figure 9.16, with bond

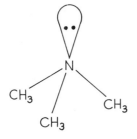

Figure 9.16. The structure of the trimethylamine molecule showing the position of the lone pair of electrons

angles slightly less than those in a regular tetrahedron (cf. the ammonia molecule). The methyl group may be regarded as methane less one hydrogen atom so that the singly occupied sp^3 hybrid orbital is used in the bonding to the nitrogen atom. The nitrogen atom possesses a lone pair of electrons in an sp^3 hybrid orbital.

An identical bonding scheme could operate in trisilylamine, $N(SiH_3)_3$, and it could be expected that this molecule, like ammonia and trimethylamine, would be pyramidal. It is not. Experimental observations have shown that the three silicon atoms and the nitrogen atom are coplanar. This fact may be explained in terms of p_π—d_π bonding occurring between the atoms involved. This is not possible with carbon but the silicon atom possesses low lying d orbitals which may be used in trisilylamine. If we take the lone pair in nitrogen to be accommodated in the p_z orbital instead of an sp^3 hybrid orbital, then this may overlap with the d_{xz} orbitals of the silicon atoms, as shown in Figure 9.17, to give a p_π—d_π system of a bonding orbital, two non-bonding orbitals and an antibonding orbital. Only two electrons are available and they will normally occupy the bonding orbital. The N—Si bonds will have a bond order of $1\frac{1}{3}$ due to the p_π—d_π bonding. The electrons in this bonding orbital are both supplied

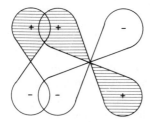

Figure 9.17. The overlap of the p_z orbital of a nitrogen atom with the d_{xz} orbital of a silicon atom in the trisilylamine molecule

Other Polyatomic Molecular Systems

by the nitrogen atom and as such the bond may be said to be a coordinate bond. The coplanarity of the nitrogen and silicon atoms is essential for the bonding scheme to operate and, of course, the N—Si bonds are stronger than they would be in a pyramidal molecule. The possibility of such bonding must always be taken into account in the prediction of the shapes of molecules since, in a few cases, its operation does affect the shape.

The tendency of the lone pair electrons in trimethylamine to be donated in coordinate bond formation is exemplified by its reaction with boron trifluoride,

$$(H_3C)_3N + BF_3 \rightarrow (H_3C)_3N—BF_3$$

in which the bonding around both the nitrogen and boron atoms becomes tetrahedral. The bond between the boron and nitrogen atoms is described as a coordinate bond since both electrons are supplied by one of the atoms (nitrogen in this case).

Trisilylamine has very little tendency to form this kind of bond with an *acceptor* molecule such as BF_3, presumably because of the added stability which the intramolecular coordinate bond system confers upon the electron pair which would be used in such compound formation.

9.5 Some Inter-related Polyatomic Molecules

The AB_n type molecule has been treated in general, and now several examples will be dealt with of molecules having a slightly more complex structure in that they possess more than one 'central' atom. The molecules to be considered here fall into three isoelectronic groups. They are:

Group 1. ethane, C_2H_6, hydrazine, N_2H_4, hydrogen peroxide, H_2O_2.
Group 2. ethylene, C_2H_4, formaldehyde, CH_2O.
Group 3. acetylene, C_2H_2, hydrogen cyanide, HCN.

Group 1

(a) *Ethane* C_2H_6

This molecule may be thought to be formed by the joining together of two methyl radicals (CH_3) by a single C—C bond. Methane is a perfectly tetrahedral molecule, the removal of a hydrogen atom from which produces the methyl radical, CH_3. We may consider that the tetrahedral angles of 109° 28' are preserved (although in isolation as these would have different values). The single electrons in the sp^3 hybrid orbitals of the two carbon atoms may then be used to form the C—C sp^3—sp^3 bond. Surrounding each carbon atom in ethane there would then be four bonding pairs of electrons (three C—H and one C—C) which would be expected to give a near-tetrahedral distribution of the bonds in space.

Bearing in mind the principles of the minimization of electron pair repulsion, we may now consider the relative dispositions of the two 'ends' of the ethane molecule. There are two extreme conformations possible, these being illustrated by end-on views of the molecule (i.e. viewed down the C—C axis) as shown in Figure 9.18. These conformations may be called staggered and eclipsed respectively. The staggered conformation (a) is the one in which the repulsions between the bonding pairs in the two methyl groups are minimized. They are at a maximum in the eclipsed conformation (b).

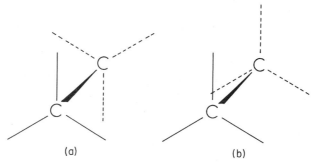

Figure 9.18. The staggered (a) and eclipsed (b) conformations of the ethane molecule

Apart from such considerations there is no restriction to the free rotation of one methyl group with respect to the other one around the C—C bond which has the cylindrical symmetry typical of a sigma bond. In such a rotation the potential energy would pass through a maximum when the molecule was in position (b) and a minimum when position (a) was attained. The difference in energy between these two positions is known as the barrier to free rotation which, in the case being considered, is estimated to be 3 kcal.mole^{-1} (12·6 kJ.mole^{-1}). This barrier to rotation hinders the process which is not strictly 'free'. The barrier in the case of ethane is small so that hindered rotation about the single bond does occur even at room temperatures.

(b) *Hydrazine*, N_2H_4

This molecule may be constructed by removing a hydrogen atom from each of two ammonia molecules and by forming an sp^3-sp^3 N—N sigma bond, with the steric arrangements around each nitrogen atom remaining virtually as they were in the original ammonia molecules. Each nitrogen

Other Polyatomic Molecular Systems

atom is surrounded by four pairs of electrons (two N—H, one N—N and one non-bonding) and these will have a non-regular tetrahedral spatial distribution. As was the case with the ethane molecule, there are possibilities of restricted or hindered rotation. The presence of the two non-bonding pairs and their greater repelling characteristics give rise to the different extreme conformations of the molecule as shown in Figure 9.19. The molecule has a dipole moment of 1·85D (1D = 10^{-18} e.s.u. = $3·33 \times 10^{-26}$C) and this rules out the *trans* form (b), which would have a zero dipole moment. There is no evidence for the eclipsed form (a), and the equilibrium conformation is probably the *gauche* form shown in Figure 9.19(c).

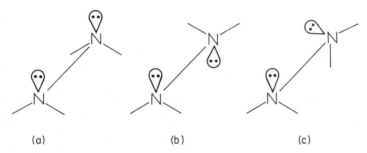

Figure 9.19. The *cis* (a), *trans* (b) and *gauche* (c) conformations of the hydrazine molecule

(c) *Hydrogen Peroxide*, H_2O_2

In this molecule the oxygen atoms are surrounded by one H—O bonding pair, one O—O bonding pair and two non-bonding pairs of electrons. These will be arranged in a tetrahedral manner and, in practice, the molecule is found to have the equilibrium form shown in Figure 9.20.

In the molecules, C_2H_6, N_2H_4 and H_2O_2 (which are isoelectronic with F_2), there are the units, CH_3, NH_2 and OH, which are isoelectronic with the fluorine atom. It is of interest to note that simple molecular orbital theory predicts a bond order of unity for the F—F linkage and would make this same prediction for the C—C, N—N and O—O bonds in the above molecules. Some experimental evidence may be quoted to support such a prediction. The data in Table 9.4 include the lengths and energies of such bonds. The bonds are of similar length and the bond strengths are of the same order (when compared with those of multiple bonds). The decrease in bond dissociation energy in going from C—C to F—F (Table 9.4) may be associated with the increase in lone pair–lone pair repulsive effects.

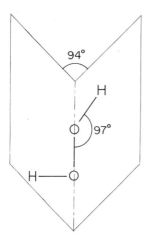

Figure 9.20. The geometry of the hydrogen peroxide molecule

Table 9.4. Bond Length and Strength Data for Some Single Bonds

Bond	Length (nm)	Bond dissociation energy	
		(kcal.mole^{-1})	(kJ.mole^{-1})
C—C in ethane	0·154	83·0	347
N—N in hydrazine	0·147	66·0	276
O—O in hydrogen peroxide	0·149	48·0	201
F—F in fluorine	0·142	37·7	158

Group 2

(a) *Ethylene,* C_2H_4

Experimentally this molecule is found to be planar, the bond angles being 120°. The sigma bonding may be described in terms of sp^2 hybridization of the carbon orbitals leaving one pure p orbital on each carbon to form the pi bond between the carbon atoms. The C—C sp^2—sp^2 sigma bond would be subjected to hindered rotation as discussed above for ethane. However, the formation of the pi bond removes the relevance of such a discussion in the case of ethylene since the carbon p orbitals overlap in such a manner as to prevent rotation about the C—C bond unless the pi link is first of all broken. The pi bonding ensures the coplanarity of all six atoms. The fixing of the ethylene structure makes possible the phenomena of the *isomerism* of disubstituted compounds, $C_2H_2A_2$. Such compounds may exist in all of the isomeric forms shown in Figure 9.21.

Other Polyatomic Molecular Systems

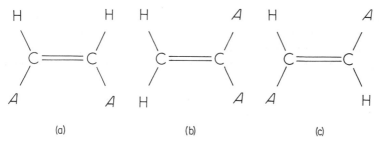

(a)   (b)   (c)

Figure 9.21. The three isomeric forms of disubstituted ethylene molecules: (a) *cis*, (b) *geminal*, and (c) *trans*

Linnett has suggested an alternative approach to double bonds which does not involve sigma and pi distinctions. Just as the single C—C bond in ethane may be thought of as being formed by two tetrahedra sharing a common vertex as in Figure 9.22, the double bond in ethylene may be

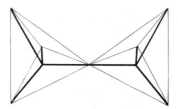

Figure 9.22. Representation of the ethane molecule by two tetrahedra sharing a common vertex

constructed on the basis of the sharing of a common edge between two such tetrahedra as in Figure 9.23. To reconcile such a formulation of the

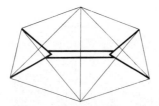

Figure 9.23. Representation of the ethylene molecule by two tetrahedra sharing a common edge

double bond in ethylene with the orbital concepts outlined in this book, it is necessary to view the bonding as being composed from the overlapping of two sp^3 hybrid orbitals from each of the carbon atoms. The sp^3—sp^3 bonds would be 'bent' as is shown in Figure 9.24. Such bent

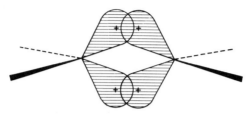

Figure 9.24. A representation of the formation of two 'bent' sp^3–sp^3 bonds necessary to the description of the ethylene molecule by two tetrahedra sharing a common edge

bonds are not normally thought to be common, but have been postulated to explain the tetrahedral P_4 molecule which has $P\widehat{P}P$ bond angles of 60°. Such bond angles cannot be formed on the basis of the normal hybridization schemes and hence the use of the bent bond. Another example of a bent bond is used to explain the structure of the $Co_2(CO)_8$ molecule. This may be built up from two 'octahedra' such as is shown in Figure 9.25, in which three of the octahedral positions are occupied by carbon monoxide molecules which form coordinate bonds with the central

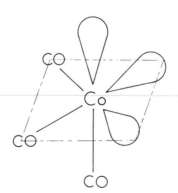

Figure 9.25. The octahedral arrangement of three CO ligands and three lone pairs of electrons around a central cobalt atom

Other Polyatomic Molecular Systems

cobalt atom. When two such octahedra are joined up using two carbonyl bridges we get the observed structure, as shown in Figure 9.26, in which there is a bent bridging bond. This is in order to conserve the octahedral distribution around the cobalt atoms which is found to occur in many transition metal compounds, details of which are treated in Chapter 11.

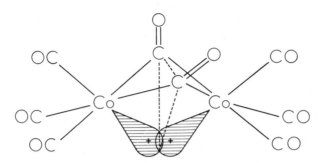

Figure 9.26. The structure of $Co_2(CO)_8$ involving two carbonyl (C=O) bridging groups and a bent bond

(b) Formaldehyde

The ethylene molecule is isoelectronic with the oxygen molecule and there are similarities between the two in terms of bond lengths, energies and orders. The CH_2 group and the oxygen atom are also isoelectronic and the formaldehyde molecule, CH_2O, which contains both these species is expected to show similarities to both ethylene and oxygen. The bonding in formaldehyde may be described in terms of sp^2 hybridization of the carbon orbitals (as in ethylene), to give a trigonally planar distribution of orbitals used in the sigma bonding. The p orbital perpendicular to the trigonal plane is used in the pi bonding to the oxygen atom. This ensures coplanarity of the atoms involved.

The data for the comparison of the bonds of order *two* in the molecules, C_2H_4, CH_2O and O_2, are presented in Table 9.5.

Table 9.5. Bond Strength and Length Data for Some Double Bonds

Bond	Molecule	Length (nm)	Bond dissociation energy	
			(kcal.mole^{-1})	(kJ.mole^{-1})
C=C	C_2H_4	0·134	148	620
C=O	CH_2O	0·123	170	710
O=O	O_2	0·121	118	494

The bond lengths are considerably shorter than the single bond lengths (Table 9.4) and the energies are higher than for the single bonds. The evidence supports the description of such bonds as of order two—or at least a bond order higher than unity.

The molecule, HN=NH (diimine), which is related to hydrazine in a similar manner as ethylene is related to ethane, has not been prepared. Its derivative, azobenzene, $C_6H_5N=NH_6H_5$, is known, and, as we would expect from there being four pairs of electrons surrounding each nitrogen atom with one of them involved in the pi bond (N=N), the C—N—N—C group of atoms in the centre is not linear. There are two bonding pairs of electrons and a lone pair in the valency shell of the nitrogen atom. We would expect sp^2 type bonding so that the possible arrangements would be the *cis* and *trans* forms of the molecule as are shown in Figure 9.27.

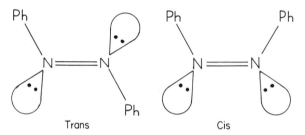

Trans   Cis

Figure 9.27. The *trans* and *cis* isomeric forms of azobenzene

In practice it is found that the compound contains molecules in the *trans* form in which all the atoms are coplanar. Further discussion of this molecule appears in Section 9.7.

Group 3

In this group we may deal with three molecules which are isoelectronic, namely, acetylene, C_2H_2, nitrogen, N_2, and hydrogen cyanide, HCN. The CH group and the nitrogen atom are also isoelectronic.

The acetylene molecule may be described in two ways. The C—C bond may be regarded as being formed by overlapping *sp* hybrid orbitals of the carbon atoms. The other *sp* hybrid orbitals may then be used in the formation of the two C—H bonds. The four atoms are, by this method of bonding, constrained to be colinear, as is shown in the overlap diagram in Figure 9.28. The two pairs of pure *p* orbitals may then overlap to give two degenerate pi bonds which are mutually perpendicular. Such a 'triple' bond is predicted by molecular orbital theory for the nitrogen molecule, and Table 9.6 shows the similarities in bond lengths and

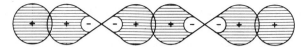

Figure 9.28. The sigma bonding in the acetylene molecule

energies between the two molecules and also for the isoelectronic HCN molecule. The bonding in the latter molecule may also be interpreted in terms of a sigma and two pi bonds. The energies and lengths of the bonds are those expected for bonds of order three.

Table 9.6. Bond Strengths and Lengths for Some Triple Bonds

Bond	Molecule	Length (nm)	Bond dissociation energy	
			(kcal.mole^{-1})	(kJ.mole^{-1})
C≡C	HCCH	0·120	194	810
N≡N	N$_2$	0·110	225	940
C≡N	HCN	0·116	210	878

The alternative way of regarding the bonding in these three molecules is in terms of the carbon and/or nitrogen atoms using sp^3 hybridized orbitals to form three 'bent' sp^3-sp^3 σ-type bonds. These situations would be summarized by the sharing of *faces* between two tetrahedra.

9.6 Some Bridged Molecules

Three molecules will be dealt with in this section, these being aluminium bromide which in the gaseous state has the composition Al$_2$Br$_6$, diborane, B$_2$H$_6$, and dialuminium hexamethyl, Al$_2$(CH$_3$)$_6$. In these molecules the two 'central' atoms are linked together by two bridging atoms or groups and have the general stereochemistry as shown in Figure 9.29. The bonds

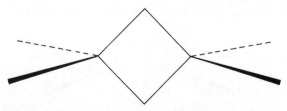

Figure 9.29. The general stereochemistry of doubly bridged molecules, A_2B_6

surrounding the central atoms are arranged in a distorted tetrahedral fashion and vary somewhat with the individual molecules.

Dialuminium Hexabromide, Al_2Br_6

The $AlBr_3$ unit, with six electrons in the valency group of the aluminium atom and with the possibility of pi bonding as in BF_3 (Section 9.1), is expected to be planar. Such a unit would also be expected to be an electron pair acceptor and in the dimer, Al_2Br_6, such donor–acceptor properties are exhibited with the aluminium atoms acting as acceptors and two of the bromine atoms (bridging) acting as electron pair donors in the formation of two coordinate links. The geometry of the molecule is shown in Figure 9.30.

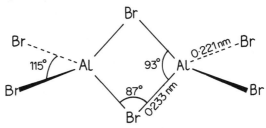

Figure 9.30. The structure of the Al_2Br_6 molecule

With such an arrangement there are then four electron pairs surrounding each aluminium atom and the spatial arrangement is roughly tetrahedral. The Al—Br—Al angle is observed to be 87°. This is consistent with there being slightly 'bent' pure *p* bonding on the part of the bromine atoms. The idea of there being four pairs of electrons surrounding the bromine atoms would indicate bond angles which were approximately tetrahedral. This is obviously not so in this particular case.

Diborane, B_2H_6

The detailed stereochemistry of the diborane molecule is presented in Figure 9.31. Unlike bromine the hydrogen atom possesses only one

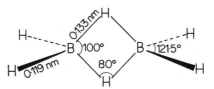

Figure 9.31. The structure of the diborane molecule

electron and cannot therefore take part in a bonding scheme such as the one postulated for aluminium bromide. In the case of diborane the bridge bonding may be interpreted in terms of a three centre molecular orbital system which makes use of an sp^3 hybrid orbital from each of the boron atoms, together with the $1s$ orbital of the hydrogen atom. The overlap diagram and the bonding molecular orbital system are shown in Figure 9.32. Two such systems complete the bridging between the two

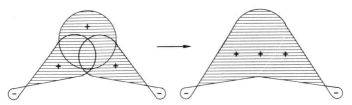

Figure 9.32. The overlap of sp^3 hybrid orbitals of two boron atoms with the $1s$ orbital of a hydrogen atom in the formation of a bonding three-centre molecular orbital

boron atoms, each of the bridge bonds containing two electrons in the bonding orbital. The non-bonding and antibonding molecular orbitals are vacant in the ground state of the molecule. The positions of the hydrogen atoms in the bridging structure are determined by the tetrahedrally orientated sp^3 hybrid orbitals on the boron atoms and the maximization of their overlap with the $1s$ orbital of the hydrogen atoms.

If diborane is passed into an ether solution of lithium hydride, LiH, a white solid, $LiBH_4$, is precipitated. This contains the tetrahedral borohydride ion, BH_4^-, which may be thought to be produced by the reaction of diborane with hydride ions, H^-:

$$2H^- + B_2H_6 \rightarrow 2BH_4^-$$

The borohydride ion is regularly tetrahedral as would be expected from the four bonding pairs of electrons contained by the valency shell of the boron atom, and the bonding may be described in terms of the more conventional sp^3 hybrid orbital scheme.

Other reactions of diborane in which it acts as an electron pair acceptor (Lewis Acid) are with trimethylamine:

$$2(CH_3)_3N + B_2H_6 \longrightarrow 2(CH_3)_3N \rightarrow BH_3$$

in which the borine radical, BH_3, is a possible intermediate, and with ammonia. The reaction with ammonia depends upon the conditions. Excess of ammonia at low temperatures produces

$$H_2B(NH_3)_2{}^+ BH_4^-$$

both ions containing central boron atoms which are tetrahedrally bonded.

At high temperatures boron nitride, BN, is produced which is a giant molecule similar to graphite (see Section 10.4).

When the molecular ratio of ammonia to diborane is 2:1 a volatile liquid is produced which has the formula, $B_3N_3H_6$ (b.p. 55°C). This compound, borazole (or borazine), receives further treatment in Section 9.7.

Dialuminium Hexamethyl, $Al_2(CH_3)_6$

Dialuminium hexamethyl possesses the stereochemistry shown in Figure 9.33. The acute bridge angle (72°) may be explained in terms of a

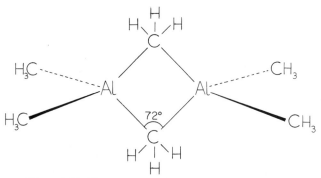

Figure 9.33. The structure of the $Al_2(CH_3)_6$ molecule

three-centre bond which involves sp^3 hybrid orbitals on all three atoms overlapping to form a molecular orbital system. The formation of the bonding orbital is shown in Figure 9.34. With three equivalent atoms the

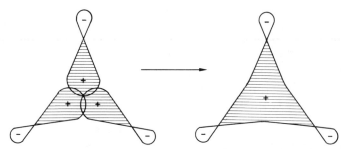

Figure 9.34. The use of three sp^3 hybrid orbitals (one from a bridging methyl group and one from each of the aluminium atoms) in the formation of a three-centre, bridge-bonding, molecular orbital

Other Polyatomic Molecular Systems

apparent bond angle would be 60°. The internuclear repulsion between the aluminium atoms ($Z = 13$), being greater than that between aluminium and carbon, produces a bond angle, Al—C—Al, greater than 60°.

9.7 Some Organic Systems: Pi Electron Delocalization

Several examples have been given of uses of triply-centred bonds and in this section such ideas will be developed to cover more complex systems where it is not possible (or may be incorrect) to describe electronic configurations in terms of localized electron pair bonds. The systems to be covered include:

        (1) butadiene,
        (2) polyenes,
        (3) benzene and borazine,
        (4) azobenzene.

Butadiene, $CH_2.CH.CH.CH_2$

The electronic arrangement in butadiene may be written in the form:

$$\overset{1}{C}H_2=\overset{2}{C}H-\overset{3}{C}H=\overset{4}{C}H_2$$

indicating that there are four carbon atoms (numbered for future reference) linked by localized sigma bonds, and that the two pairs of carbon atoms possess a pi bond each. Such a formulation, although it does not violate any principles of valence so far dealt with, is inaccurate. Justification for such a statement lies in evidence from very varied sources which include the following points:

(1) The addition of bromine to butadiene may occur at the site of one of the localized double bonds in which case the product is $BrCH_2-CHBr-CH=CH_2$ (1,2-dibromo-but-3-ene), but experimentally the major product is the one produced by the 1:4 addition of the bromine to give $BrCH_2-CH=CH-CH_2Br$ (1,4-dibromo-but-2-ene).

(2) Butadiene possesses a transoid conformation as shown in Figure 9.35 (carbon skeleton only). The cisoid conformation probably exists in the vapour phase and as an intermediate in the reactions of butadiene. Metal complexes exist in which butadiene acts as a ligand and in which the butadiene has a cisoid conformation.

(3) The bond lengths expected for C—C bonds of orders one and two respectively are 0·154 nm and 0·134 nm. The bond lengths in butadiene are as shown in Figure 9.35.

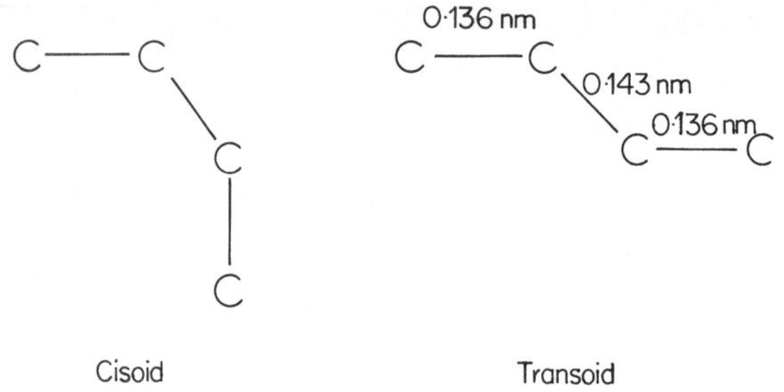

```
    C———C              C———C
         \                  \0·143 nm
          C                  C———C
          |                   0·136 nm
          C

      Cisoid              Transoid
```

Figure 9.35. The cisoid and transoid conformations of the butadiene molecule

(4) Ethylene undergoes a $\pi \to \pi^*$ transition at a wavelength of about 180 nm. In other words a quantum corresponding to a wavelength of 180 nm when absorbed by an ethylene molecule causes one of the electrons in the π bonding orbital to be excited to the otherwise vacant π^* antibonding orbital. A similar transition in butadiene is bathochromically shifted (i.e. occurs at longer wavelength) to 217 nm.

These experimental observations are at variance with the localized double bond formulation of butadiene. All the observations may be explained on the basis of some interaction between the two pi electron systems. A satisfactory pi electron system may be built upon the carbon skeleton formed by using sp^2 hybrid carbon orbitals and leaving one p orbital upon each carbon atom to form a four-centre system of π-type molecular orbitals. The four p atomic orbitals combine as shown in Figure 9.36 to produce four pi molecular orbitals. They are shown in order of their respective stabilities, ψ_1 being the lowest bonding molecular orbital which is bonding between all four carbon atoms. The next lowest orbital, ψ_2, is bonding between the two end pairs of carbon atoms but is antibonding with respect to the inner carbon atoms. The third orbital, ψ_3, is the reverse of ψ_2, and ψ_4 is completely antibonding. Numbering the carbon atoms from left to right, the characters of the four molecular orbitals, ψ_1 to ψ_4, are shown in Table 9.7.

Ignoring the effect on orbital energies of the overlap integral, the energies of the four orbitals may be represented as in Figure 9.37. The four electrons which have to be accommodated in the pi system in butadiene will, in the ground state, produce the configuration, $(\psi_1)^2(\psi_2)^2$, with ψ_3 and ψ_4 remaining unoccupied. The orbital scheme derived from the four p orbitals as described above explains immediately point (2). The scheme depends on the p orbitals (and consequently the molecular orbitals) all

Other Polyatomic Molecular Systems

Table 9.7. Pi Orbital Characters in Butadiene

Orbital	Between atoms			Overall
	1–2	2–3	3–4	
ψ_4	A^a	A	A	$3A$
ψ_3	A	B	A	A
ψ_2	B^b	A	B	B
ψ_1	B	B	B	$3B$

[a] A represents antibonding character.
[b] B represents bonding character.

being codirectional. This limits the carbon skeleton to the two isomers shown in Figure 9.35.

It also explains (but not quantitatively) point (3). The presence of pi electrons in orbitals, ψ_1 and ψ_2, causes shortening of the bonds compared

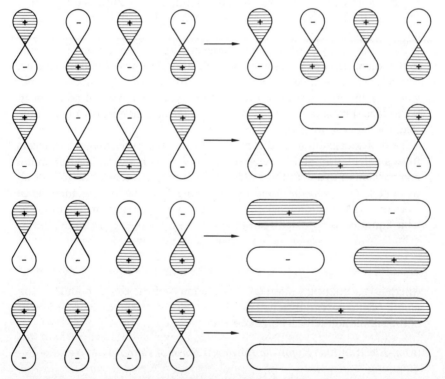

Figure 9.36. The pi orbital overlap diagrams for four p orbitals and the diagrams for the molecular orbitals thereby produced

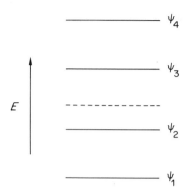

Figure 9.37. The approximate relative energies of the pi molecular orbitals of butadiene

to the single sigma C—C distance, 0·154 nm, and the antibonding nature of ψ_2 as far as atoms 2 and 3 are concerned explains the greater length of the central bond.

Point (1), which is concerned with the preferred 1:4 addition of bromine to butadiene, may be explained on the basis of the above bonding scheme, which effectually delocalizes the pi electrons over the molecule and, whatever the mechanism of the reaction, would enable 1:4 addition to take place. Probably the 1:4 product is thermodynamically more stable than the less symmetrical 1:2 product.

That delocalization of the pi electrons is probable may be deduced from a calculation of the π-electron energy of such a configuration and a comparison of this with the π-electron energy in the case of two localized pi bonds. This may be done by using the formula for the energies of electrons in a one-dimensional potential well (derived in Section 4.5). This equation is repeated below:

$$E = \frac{n^2 h^2}{8ml^2} \tag{9.1}$$

where n is a quantum number, m is the mass of the electron and l is the length of the potential well. The one-dimension in the case of butadiene is the sigma bonded carbon skeleton.

(a) *π-Electron Energy for the Localized Form of Butadiene*

$$CH_2=CH-CH=CH_2$$

The two non-interacting pi systems may be dealt with separately as

Other Polyatomic Molecular Systems

potential wells of length b (where b is one bond length). This modifies the equation (9.1) to read:

$$E = \frac{n^2 h^2}{8mb^2} \qquad (9.2)$$

The lowest energy level is that for which $n = 1$, so we may write

$$E_1 = \frac{h^2}{8mb^2} \qquad (9.3)$$

and the next lowest level is that for which $n = 2$ and has an energy:

$$E_2 = \frac{4h^2}{8mb^2} \qquad (9.4)$$

In each pi bond there are two electrons and these will be accommodated in the lowest vacant level, this being E_1 in both cases.

The total π-electron energy of the localized system will therefore be given by $4E_1$ which is equal to $h^2/2mb^2$. This we may call $E_\text{localized}$.

(b) *π-Electron Energy for the Delocalized Form of Butadiene*

Considering the pi electrons to exist in orbitals which are delocalized over the four carbon atoms, we may put the length of the single potential well equal to $3b$. As has been pointed out, the bonds in butadiene are not of equal length, but to assume that they are of equal length is a minor approximation compared to the larger one of treating butadiene as a one-dimensional potential well.

In this case the four pi electrons must occupy the two orbitals of lowest energy, these being E'_1 and E'_2 where

$$E'_1 = \frac{h^2}{72mb^2} \qquad (9.5)$$

and

$$E'_2 = \frac{4h^2}{72mb^2} \qquad (9.6)$$

The total π-electron energy is given by $2E'_1 + 2E'_2$ which we will call $E_\text{delocalized}$.

$$E_\text{delocalized} = \frac{10h^2}{72mb^2} \qquad (9.7)$$

Comparing $E_\text{localized}(h^2/2mb^2)$ with $E_\text{delocalized}(10h^2/72mb^2)$, it is clear that $E_\text{delocalized}$ is considerably smaller than $E_\text{localized}$ (by a factor of $\frac{72}{20} = 3\cdot 6$).

Since any system in its ground state exists in the lowest state of energy, we see that the π-electron delocalization which explains points (1), (2) and (3) is probable even within the limits of these approximate calculations.

Similar calculations can suggest an explanation for point (4) and give support to the general idea of π-electron delocalization.

Consider the lowest energy $\pi \to \pi^*$ transition in ethylene. The pi electrons (in level E_1) may be excited to the next level (E_2) by supplying an amount of energy, $E_2 - E_1$, to the molecule. We can use equation (9.2) to calculate the magnitude of this energy. E_1 and E_2 are given in equations (9.3) and (9.4) and the difference amounts to:

$$\Delta E = E_2 - E_1 = \frac{3h^2}{8mb^2} \tag{9.8}$$

Ethylene absorbs radiation of wavelength 180 nm which means that a transition between energy levels is taking place such that the difference in energy is equal to the energy equivalent of a quantum or photon of wavelength 180 nm. If butadiene contains two non-interacting ethylenic bonds, then we would expect that the same calculations apply and that butadiene should absorb also at 180 nm. The lowest energy transition in butadiene, however, occurs with radiation of wavelength 217 nm. The energy of this transition we can calculate to be $E'_3 - E'_2$, since E'_2 is the higher of the two occupied energy levels and E'_3 corresponds to the energy of the lowest vacant level. E'_2 is given in equation (9.6) and E'_3 by

$$E'_3 = \frac{9h^2}{72mb^2} \tag{9.9}$$

so that the transition energy is given by

$$\Delta E' = E'_3 - E'_2 = \frac{5h^2}{72mb^2} \tag{9.10}$$

Although there is not very good quantitative agreement between the calculations and the observations it is obvious (since $\frac{5}{72} < \frac{3}{8}$) that the calculations predict that delocalized butadiene should absorb at a longer wavelength than the localized or ethylenic form.

A diagrammatic representation of the lowest energy transitions in ethylene and the two forms of butadiene is shown in Figure 9.38.

Polyenes

It is of interest to carry further the ideas developed for butadiene and apply them to systems where more than four pi electrons are involved. Such systems have the general formula

$$H-(CH=CH)_n-H$$

Other Polyatomic Molecular Systems

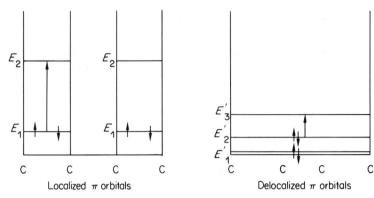

Figure 9.38. The lowest energy $\pi \to \pi^*$ transition in the localized and delocalized forms of butadiene

and are known as polyenes. Some predictions may be made concerning their absorption spectra using the electron-in-a-potential-well approximation. There are $2n$ pi electrons involved and these are contained in the lowest n energy levels in the ground states of the molecule. The energies of the highest occupied levels will be given by

$$E_n = \frac{n^2 h^2}{8mb^2} \tag{9.11}$$

Assuming, as before, that the C—C bonds are of length b, the length of the potential well in a polyene will be given by nb so that

$$E_n = \frac{n^2 h^2}{8mn^2 b^2} \tag{9.12}$$

and for reasons which will soon appear, the n^2 factors are not cancelled in equation (9.12). The energies of the lowest vacant levels will be given by

$$E_{n+1} = \frac{(n+1)^2 h^2}{8mn^2 b^2} \tag{9.13}$$

so that the lowest transition energies for polyenes are given by

$$\Delta E = E_{n+1} - E_n = \frac{(n+1)^2 h^2 - n^2 h^2}{8mn^2 b^2}$$

$$= \frac{(2n+1)h^2}{8mn^2 b^2} \tag{9.14}$$

This equation predicts that as n increases the transition energy, ΔE, decreases, i.e. as the π-electron system becomes longer the polyenes should absorb radiation at longer and longer wavelengths. This is found to be the case as is shown in Table 9.8 which lists the absorption maxima for a series of polyenes. The stereochemistry of polyene molecules is very much determined by the π-electron arrangements which 'freeze' the molecules into either cisoid or transoid conformations.

Table 9.8. Absorption Maxima for Some Polyenes

Number of π-electron pairs (n)	λ_{max} (nm)
2	220
3	270
4	310
5	340
6	380
7	400
8	420
9	440

Benzene, C_6H_6

The benzene molecule, C_6H_6, possesses a regular hexagonal planar structure as shown in Figure 9.39. The six carbon atoms may be bonded by sigma bonds which themselves are constructed from the sp^2 hybridized orbitals of the atoms. If this is the case, there are then six p orbitals (one on each carbon atom) which are all equivalent and which can overlap to form a pi orbital system which is delocalized over the six carbon atoms.

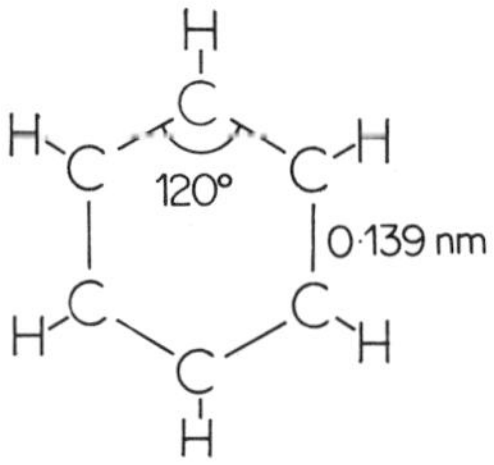

Figure 9.39. The structure of the benzene molecule

That this is the case is clear from various aspects of experimental observation. A molecule involving three localized pi bonds (ethylenic) would have properties typical of such bonds. Benzene undergoes many *substitution* reactions as well as the addition reactions expected for an

Other Polyatomic Molecular Systems

unsaturated molecule. For example, when benzene is treated with a mixture of concentrated nitric and sulphuric acids a nitro group ($-NO_2$) is substituted for a hydrogen atom with the production of nitrobenzene, $C_6H_5NO_2$.

The molecule containing three localized double bonds (cyclo-hexa-1,3,5-triene) would not be expected to be regularly hexagonal with a C_6 axis of symmetry but would only possess a C_3 axis as shown in Figure 9.40, since double bonds are shorter than single bonds. Furthermore, benzene absorbs radiation at 256 nm—a much longer wavelength than does ethylene or systems containing isolated (localized) double bonds (180 nm).

Consideration of the symmetry of the benzene molecule shows that the six π-type orbitals formed from the six p atomic orbitals are formed by the overlaps shown in Figure 9.41. If we number the atoms in the benzene molecule as in Figure 9.42, we can summarize the character of each of the orbitals, ψ_1 to ψ_6, in terms of their bonding or antibonding properties between each pair of adjacent atoms. This character summary is shown in Table 9.9.

The approximate relative energies of these orbitals are shown in Figure 9.43. ψ_1 is the lowest in energy and is bonding with respect to all six carbon atoms. ψ_2 and ψ_3 have overall bonding character but due to their antibonding or non-bonding components, they are of higher energy than ψ_1. They form a doubly degenerate energy level. The photoelectron spectrum of benzene indicates that ionization of an electron from the highest filled level (ψ_2, ψ_3) requires 9·4 ev of energy. The totally bonding orbital, ψ_1, is 2·4 ev.atom^{-1} (232 kJ.mole^{-1}) more stable than the degenerate ψ_2, ψ_3 level. The other doubly degenerate orbitals, ψ_4 and ψ_5, are overall antibonding and the ψ_6 orbital is completely antibonding. The six pi electrons in the ground state of the molecule would have the configuration, $(\psi_1)^2(\psi_2)^2(\psi_3)^2$. The lowest energy transition would be from the ψ_2, ψ_3 level to the vacant ψ_4, ψ_5 level.

As was the case with the polyenes, the greater the extent of the delocalization (the larger the l value of the potential well) the lower the energy (longer the wavelength) of the lowest energy transition. The low energy of the ground state π-electron configuration and its fairly high symmetry are two factors responsible for the relative (i.e. to ethylenic compounds) inertness to addition reactions. An addition reaction would have the effect of destroying the high symmetry of the π-electron configuration and would result in a considerable loss of what we might call the *delocalization energy*. Substitution reactions are preferred in which the π-electron configuration is preserved.

Figure 9.40. The structure and bonding of the cyclo-hexa-1,3,5-triene molecule showing alternating bond lengths and the C_3 axis (perpendicular to the molecular plane)

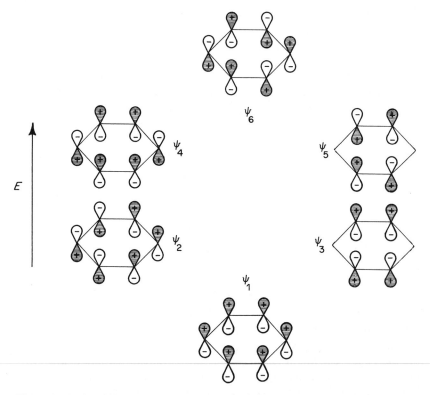

Figure 9.41. Overlap diagrams for the six pi molecular orbitals of the benzene molecule

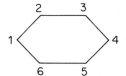

Figure 9.42. The numbering of the carbon atoms of the benzene ring as referred to in Table 9.9

Other Polyatomic Molecular Systems

Table 9.9. Pi Orbital Characters for the Benzene Molecule

Orbital	Character between atoms						Overall
	1–2	2–3	3–4	4–5	5–6	6–1	
ψ_6	A^a	A	A	A	A	A	$6A$
ψ_5	—	A	—	—	A	—	$2A$
ψ_4	A	B	A	A	B	A	$2A$
ψ_3	—	B	—	—	B	—	$2B$
ψ_2	B^b	A	B	B	A	B	$2B$
ψ_1	B	B	B	B	B	B	$6B$

[a] A represents antibonding character.
[b] B represents bonding character.

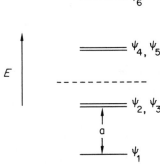

Figure 9.43. The approximate relative energies of the pi molecular orbitals of the benzene molecule, where $a = 2.4$ ev.atom^{-1}, 55.3 kcal.mole^{-1}, 232 kJ.mole^{-1}

The delocalization energy is the amount by which the delocalized form of the molecule is more stable than that involving localized double bonds. In the case of butadiene this energy is given by the difference, $4E_1 - (2E'_1 + 2E'_2)$, (Section 9.7).

Similar calculations can be made for benzene but an alternative estimation will be made here which is based on experiment. The molecule cyclo-hexene, whose structure is shown in Figure 9.44, contains one double bond and may serve as the basis of the estimation of the delocalization energy of benzene. The heat of hydrogenation of cyclo-hexene is -28.6 kcal.mole^{-1} (-120 kJ.mole^{-1}). We may make the reasonable assumption that the cyclo-hexatriene molecule should have a heat of hydrogenation equal to three times this value, i.e. $3 \times -28.6 = -85.8$ kcal.mole^{-1}

Figure 9.44. The cyclo-hexene molecule

(-360 kJ.mole^{-1}). The heat of hydrogenation of benzene is observed to be $-49\cdot8$ kcal.mole^{-1} (-209 kJ.mole^{-1}). The significance of these figures is that, taking the product of the hydrogenation reactions (cyclo-hexane, C_6H_{12}) as the reference standard, benzene is 36 kcal.mole^{-1} (151 kJ.mole^{-1}) more stable than cyclo-hexatriene. This situation is shown in Figure 9.45.

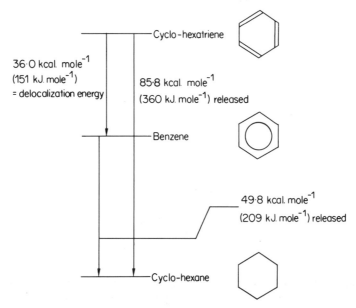

Figure 9.45. The energies of the benzene and cyclo-hexatriene molecules relative to that of cyclo-hexane

Another point deserving of consideration is that the bond lengths (C—C) in cyclo-hexene are 0·133 nm and 0·150 nm respectively for the double and single bonds and that the calculation of the delocalization energy of benzene is based on the hydrogenation energy of three bonds (localized) of length 0·133 nm compared to the bonds in benzene which are *all* 0·139 nm long. A more accurate estimate would take into account this difference in bond lengths. The localized double bond of length 0·133 nm is in its most stable state at this length. Vibration of the two

Other Polyatomic Molecular Systems

carbon atoms participating in such a bond would cause deviations from this equilibrium distance and whichever way (extension or contraction) this deviation occurs, the system will move to a higher potential energy. From a study of the vibrations of molecules (infrared and Raman spectroscopy) the energy required to stretch a double bond from a length of 0·133 nm to one of 0·139 nm and to compress a single bond from 0·150 nm to 0·139 nm may be estimated. To equalize the bond lengths in cyclo-hexatriene requires 27 kcal.mole^{-1} (113 kJ.mole^{-1}) and this has to be added to the thermochemically estimated delocalization energy of 36 kcal.mole^{-1} (151 kJ.mole^{-1}) to give the more realistic value of 63 kcal.mole^{-1} (264 kJ.mole^{-1}) for the delocalization energy. This situation is represented in Figure 9.46.

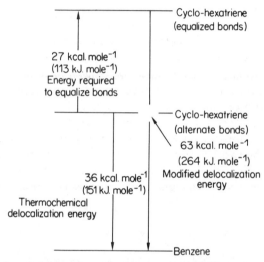

Figure 9.46. A comparison of the energies of the two hypothetical forms of cyclo-hexatriene (one with equalized bonds, the other with alternating bonds) and the benzene molecule

The $B_3N_3H_6$ molecule known as borazole or borazine is isoelectronic with benzene. The molecule is hexagonal with alternate BH and NH groups. The six electrons which form a delocalized system similar to that of the benzene molecule are supplied by the three nitrogen atoms. Nitrogen is more electronegative than boron and the electron distribution in borazine is polarized with the electron density in the vicinity of the nitrogen atoms being greater than that near the boron atoms. This polarization causes differences in properties of borazine as compared to benzene. In particular it aids addition reactions although borazine takes part in some substitution reactions as well.

Azobenzene, $C_6H_5N_2C_6H_5$

The molecule of azobenzene has a planar *trans* conformation as shown in Figure 9.47 in which the delocalized pi electrons in the benzene rings

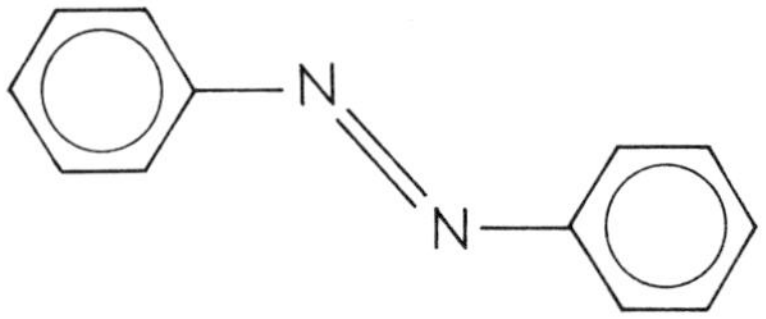

Figure 9.47. The *trans*-azobenzene molecule. The benzene rings are symbolized in the conventional manner

are symbolized in what is now the conventional manner. In Figure 9.47 the benzene rings are shown to have delocalized pi systems and there is an isolated double bond shown between the nitrogen atoms. This is only a formal description however and the coplanarity of the benzene rings together with evidence from the absorption spectrum (azobenzene is red) is good evidence for the participation of the nitrogen *p* orbitals and the pi systems of the two benzene rings in an extensive delocalization of the pi electrons. This is broken down when the compound is reduced to hydrazobenzene (shown in Figure 9.48) which is colourless and in which

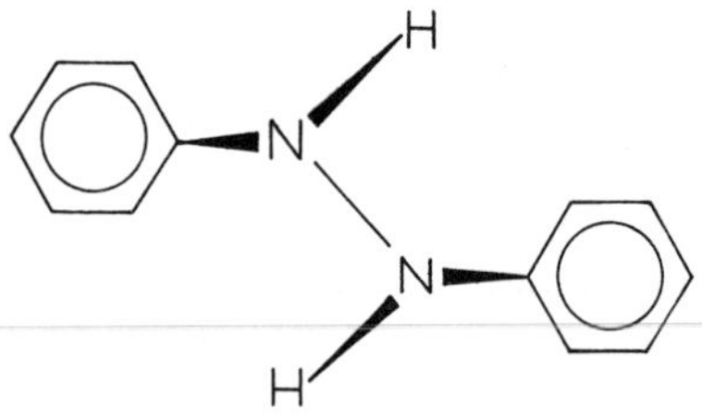

Figure 9.48. The structure of the hydrazobenzene molecule

there is a pyramidal arrangement around each nitrogen atom (sp^3 hybridization) as in the ammonia molecule (Section 9.2).

It is convenient to represent many molecular systems by using single–double bond nomenclature, but it must be realized that wherever possible such formally represented pi systems will be delocalized as the chemical, stereochemical and spectroscopic evidence indicates.

9.8 Valence Bond Theory

The treatment of molecules so far has been dependent mainly upon the Molecular Orbital Theory which makes use of the Linear Combination of Atomic Orbitals (LCAO). It is fairly generally applicable to molecular systems and is successful in a qualitative (and sometimes quantitative) manner in the explanation and interpretation of their properties.

Historically it was preceded in many cases by the Valence Bond Theory which is still extremely useful in the interpretation of the properties of organic molecules in particular.

This theory depends upon the contribution to the total molecular wave function of wave functions representing what are known as *canonical forms*. In the case of benzene two of these canonical forms are shown in Figure 9.49 and are known as the 'Kekule' structures. It should be appreciated that these canonical forms of benzene are the planar regularly

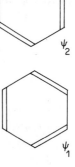

Figure 9.49. The two 'Kekule' canonical forms of the benzene molecule

hexagonal forms of cyclo-hexatriene, i.e. those with equalized bond distances, and as such have no normal existence (since they are 27 kcal.mole^{-1} (113 kJ.mole^{-1}) above the ground state which has alternating bond lengths (Section 9.7)).

The total molecular wave function consists of a linear combination of the wave functions for the separate canonical forms, ψ_1 and ψ_2, and since they are identical may be written with unit coefficients giving

$$\psi = \psi_1 + \psi_2 \tag{9.15}$$

Thus every C—C bond is equalized, all having 50 per cent single bond character and 50 per cent double bond character.

To make the valence bond molecular wave function more accurate, the 'Dewar' forms of benzene have to be incorporated, this time with different coefficients than the 'Kekule' forms (Figure 9.49). The 'Dewar' forms are shown in Figure 9.50 which are still *planar* and have a 'long

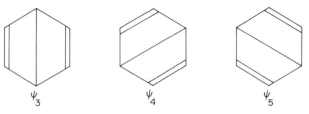

Figure 9.50. The 'Dewar' canonical forms of the benzene molecule

bond' each. With these forms incorporated the molecular wave function would be given by

$$\psi = a_1\psi_1 + a_2\psi_2 + a_3\psi_3 + a_4\psi_4 + a_5\psi_5 \qquad (9.16)$$

where $a_1 = a_2$ and $a_3 = a_4 = a_5$, but $a_1 \neq a_3$. Further refinements involving ionic forms are possible, the general picture of the π-electronic configuration given by valence bond theory being one of delocalization. The molecular orbital theory, however, gives a clearer expression of this.

For ground states of molecules both theories should give the same answer for the molecular wave function and this may be demonstrated in the case of the hydrogen molecule. The valence bond theory, however, becomes extremely complex (many canonical forms) for large molecules. The molecular orbital theory is possibly more complex to operate but eventually produces more realistic answers to problems of electronic configurations, bond orders and the spectroscopy of molecular systems.

The concept of the canonical form can lead to confused thinking about molecular systems. Such confusion derives from the use of the much maligned term 'resonance'. The canonical forms of a molecule are said to be in resonance with each other and the molecule is described as a resonance hybrid of its canonical forms. These statements must be understood *not* to mean that there is any form of physical resonance or interchange of the forms of the molecule. It does *not* mean that the molecule takes the form of one of the contributing structures for a certain percentage of its time and then electron exchange produces one of the alternative forms. Once these points are understood valence bond theory can give very satisfactory explanations of many molecular structures and of their reactions. Further examples of the application of valence bond theory or resonance theory will be mentioned in the text without any further development of the theory.

Other Polyatomic Molecular Systems

Problems

Discuss the shapes and bonding of the following molecules:

- 9.1 NF_3
- 9.2 H_2CO
- 9.3 BO_3^{3-}
- 9.4 ClO_3^-
- 9.5 $SnBr_{4(g)}$
- 9.6 SO_4^{2-}
- 9.7 ICl_4^-
- 9.8 $PCl_{5(g)}$
- 9.9 BrF_5
- 9.10 PF_6^-
- 9.11 C_3O_2
- 9.12 CH_2-C-CH_2 (allene)
- 9.13 AsH_3
- 9.14 H_3O^+
- 9.15 SeO_3^{2-}
- 9.16 SbF_5
- 9.17 BrF_3
- 9.18 TeF_6
- 9.19 $BiCl_6^{3-}$
- 9.20 $Pb(CH_3)_4$
- 9.21 HN_3
- 9.22 $C_6H_5NO_2$
- 9.23 octatetraene, $CH_2CHCHCHCHCHCH_2$
- 9.24 diphenyl, $C_6H_5-C_6H_5$
- 9.25 vinyl acetylene, $CH_2CHCHCH$
- 9.26 pyridine C_5H_6N
- 9.27 C_6H_5Cl
- 9.28 $CH_3CO_2^-$ (acetate ion)
- 9.29 HN_4^+
- 9.30 N_2O_4

10

The Solid State and the Bonding in Ionic Compounds and Metals

10.1 Introduction: Physical States of Matter

The three physical states in which we may find atomic and molecular systems are the gaseous, liquid and solid states. To a very large extent the factors which determine the physical state of any system are the *temperature* of the system and the *nature* of the *cohesive forces* which hold together the units of the system. By increasing the temperature of a solid system it is possible (at the melting point) to overcome the forces which are responsible for the cohesion of the system such that the units of the originally solid state possess translational freedom. This translational freedom is further increased at higher temperatures when the system may exist in the gaseous state. In general then, the transition from solid to liquid to the gaseous state requires energy in order to overcome the cohesive forces. The magnitudes of the melting point and boiling point of a system are a good indication of the strength of the cohesive forces operating in that system. If the forces are weak then low physical constants are to be expected and if the forces are strong the physical constants will be high.

It is possible to classify solids in terms of the *type* of cohesive force responsible for their stability as solids (i.e. the force which prevents their being liquids or gases). The three classes of solids are:

(1) *Ionic* solids, in which the primary cohesive force is the coulombic (electrostatic) force of attraction between oppositely charged *ions*, e.g. sodium chloride which contains Na^+ and Cl^- ions.
(2) *Covalent* solids, in which the cohesion is due to the covalent (electron pair) bond, e.g. diamond.
(3) *Molecular* solids, which contain distinguishable molecules (which have a separate stable existence in the gaseous state) held together in the solid by *intermolecular* or Van der Waals forces, e.g. naphthalene, iodine, carbon dioxide. In such solids the molecular units are packed

The Solid State and the Bonding in Ionic Compounds and Metals

together in such a way as to allow maximum economy of space since intermolecular forces are non-directional. The molecules in molecular solids to a large extent retain their gas phase characteristics. This is not so with ionic and covalent solids in which chemical bonds are broken in the transition to the gas phase.

With such a variety of solids as does exist and with only three general classes of solid in which to divide them it is not surprising that there are exceptions to the general rule. These exceptional cases arise because of the possibility of the cohesion being of a different type to the three classes mentioned above or to there being more than one important cohesive force in operation. For example, the vast class of minerals known as silicates consist in the main of polymeric ions in which the bonding is covalent. The negative silicate ions together with a suitable number of positive ions form the compound's *lattice* or solid arrangement. Both ionic and covalent bonding are strong and it would be difficult to place the silicate minerals in either class (1) or class (2). They should appear, if not as a separate class, in both classes (1) and (2).

Metals are another form of solid which do not easily fit into the three classes above. The bonding in metals is not fundamentally different from that in covalent molecules, but because of the special properties of metals (electrical conduction in particular) they may be treated as a separate class.

10.2 Intermolecular Forces

Helium at room temperature is a monatomic gas. Its electronic configuration ($1s^2$) is one which is stable with respect to compound formation since the excited $1s^1 2s^1$ triplet divalent state is 453 kcal.mole^{-1} (1893 kJ.mole^{-1}) above the $1s^2$ ground state. The formation of two bonds would have to cause the release of at least this amount of energy for a triatomic molecule to be formed. This is clearly not possible since the strongest single bond (H—F) so far observed is found to be 135 kcal.mole^{-1} (565 kJ.mole^{-1}). The formation of diatomic molecules is also precluded (Section 7.3) and we may assume that helium atoms do not interact with one another or any other atom in any chemical manner. Yet by cooling helium to below a temperature of $-269°$C it is possible to obtain it in the liquid state. This is an indication that there is a weak cohesive force operating which at low temperatures (i.e. low thermal agitation of atoms) can hold the helium atoms in the liquid state. The solid state can be produced at an even lower temperature provided that the pressure of helium above the solid is at least 25 atmospheres.

The weak cohesive force operating between the atoms is known as the London force. Its origin lies in the vibration of the electron cloud with respect to the nucleus. This vibration results at any one time in the atoms being instantaneous dipolar species. These may induce similar dipoles in the nearest neighbour atoms, the result being an overall attraction between the dipolar atoms. A time average of such dipoles would indicate the spherical symmetry of the atoms which have no permanent dipole moment. The London force depends on the polarizabilities of the interacting species and decreases as the inverse sixth power of the distance apart of two atoms (or molecules). This may be expressed as

$$F \propto \frac{\alpha^2}{r^6} \tag{10.1}$$

for the force of interaction between two identical species of polarizability, α. Since the polarizability of an atom or ion increases with its size, it follows that the larger members of Group 0—Ne, Ar, Kr, Xe and Rn—have progressively larger London forces of cohesion. Their boiling points and more particularly their heats of evaporation increase down the group. There is a repulsive part of this *intermolecular* force which operates at very short range and is proportional to the inverse twelfth power of the intermolecular distance.

Intermolecular forces or Van der Waals forces operate in all molecular and atomic systems. They are small in magnitude compared to the forces involved in binding atoms together to form molecules and those involved in the binding of electrons and nuclei. Apart from the London force there are two other main types of intermolecular force—the Debye orientation and Keesom induction forces.

Debye Force

This is the basis of the interaction of two dipolar species, i.e. molecules possessing permanent dipole moments, μ (see equation 7.49). The lowering of potential energy as a result of dipole–dipole orientation is proportional to μ^2/r^6 as far as the attraction is concerned.

Keesom Force

A molecule with a permanent dipole moment, μ, may induce a dipole moment, μ', in a molecule with no permanent moment, the time averaged interaction between the two being proportional to $\mu\mu'/r^6$. All three types of intermolecular force depend on the inverse sixth power of the internuclear distance.

The Solid State and the Bonding in Ionic Compounds and Metals

Under conditions of low temperature (and sometimes of high pressure) all substances become solid. In the solid phase there is generally a regular arrangement of atoms, ions or molecules which we call the crystal lattice. Under such conditions the translational energy of the constituents of the lattice is minimized and zero. Molecules may still vibrate and even rotate but only at higher temperatures do the molecules possess translational energy. Eventually with greater thermal agitation the translational capabilities of the constituents of the system increase, and the gaseous phase results. At a given temperature only cohesive forces operate against the thermal motion of the system and if they predominate a solid results, or sometimes a liquid if partial translational motion is possible.

By comparison with the Group 0 elements ionic compounds have very high melting and boiling points and high heats of fusion and evaporation. This may be understood in terms of the magnitude of the forces responsible for the cohesion of the solid materials. In an ionic substance the main cohesive force is the coulombic one which results in the production of a regular lattice. The contribution of intermolecular forces is minimal. This is not so with solids containing covalently bound molecules as, for example, naphthalene and to a very large extent all organic compounds, which have melting and boiling points far less than ionic substances of similar molecular weights. This is because in the conversion of the solid into the liquid or gaseous states the cohesive forces which have to be overcome are intermolecular. None of the covalent (strong) bonds are broken in such processes.

As was pointed out in the introduction to this chapter (Section 10.1), melting and boiling points are to a large extent governed by the magnitude of the forces responsible for the cohesion of the solid or liquid. A summary of this situation with examples is shown in Table 10.1.

Table 10.1. The Physical Constants of Some Solids and Liquids

Type of compound	Examples	(°C)	
		m.p.	b.p.
Ionic	NaCl	800	1465
	CaF_2	715	1250
Molecular	$C_{10}H_8$	80·2	218
	I_2	114	183
Covalent	Graphite	3730	4830
	Silicon	1410	2680

One other factor which may be correlated with melting point and boiling point data is the molecular weight of the molecule involved. If we consider the series, CH_4, SiH_4, GeH_4, SnH_4, of Group IV hydrides, we find that the physical constants increase with increasing molecular weight as shown in the graphs in Figures 10.1 and 10.2. Similar plots for the hydrides of Groups V, VI and VII are also shown in Figures 10.1 and 10.2.

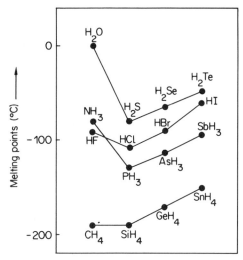

Figure 10.1. The melting points (°C) of the hydrides of elements in Groups IV, V, VI and VII

The general decrease in physical constants with decreasing molecular weight shown by all members of each series except the lightest makes possible an extrapolation to give the physical constant value expected for molecules with molecular weights corresponding to the lightest members of each series. The data in Table 10.2 show the discrepancies between the observed and extrapolated physical constants for the hydrides of the lightest elements in Groups VI, VII and VIII. The heats of fusion and evaporation of NH_3, H_2O and HF are also abnormally high compared with the other hydrides in each series.

It would appear that in the cases of NH_3, H_2O and HF there is an extra cohesive force which is responsible for the exceptionally high physical data observed. The explanation of these abnormal data depends upon the existence of *hydrogen bonding*. This is, in strength, intermediate between chemical bonding and intermolecular bonding, and is essentially an interaction of molecules containing a hydrogen atom and a very

The Solid State and the Bonding in Ionic Compounds and Metals

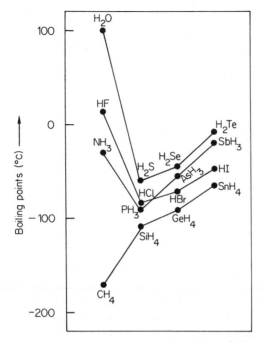

Figure 10.2. The boiling points (°C) of the hydrides of elements in Groups IV, V, VI and VII

Table 10.2. Observed and Extrapolated Values of the Melting and Boiling Points of Some Low Molecular Weight Hydrides

Hydride	(°C)			
	m.p.	Extrapolated m.p.	b.p.	Extrapolated b.p.
NH_3	−78	−140	−33	−120
H_2O	0	−95	100	−75
HF	−83	−120	19	−95

electronegative element such as nitrogen, oxygen or fluorine. The strongest hydrogen bond so far observed is in the ion HF_2^- which is a linear ion F—H—F in which the strength of the hydrogen bond is 30 kcal.mole^{-1} (126 kJ.mole^{-1}). This bond is different from that in the other systems which have been mentioned in that the hydrogen atom is symmetrically placed halfway between the fluorine atoms. It is usual in hydrogen bonding for the hydrogen atom to be nearer to one of the electronegative

244 **Atomic and Molecular Structure**

atoms than it is to the other. For instance, in solid water the oxygen atoms are tetrahedrally surrounded by hydrogen atoms, two of which are linked to the central oxygen atom by means of covalent bonds and the other two are hydrogen bonded to the oxygen in the same positions as the non-bonding orbitals are found. The situation is shown in Figure 10.3,

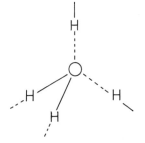

Figure 10.3. The four-coordination of the oxygen atom in the condensed phases of water

the O-----H hydrogen bonded distances being 0·178 nm and the O—H covalent bonds having a length of 0·1 nm. This structure persists throughout solid water and accounts for the effect of melting upon the specific volume. When thermal energy is supplied to a body the increased vibrational and translational energy causes an expansion of the system (at constant pressure). This is not the case for solid water. Undoubtedly the greater thermal energy of the molecules in the liquid phase would ensure an expansion when solid water melts to form liquid water. That an actual contraction occurs may be interpreted in terms of the partial destruction of the hydrogen-bonded open structure so that in the liquid molecules may pack together more closely than in the solid. The effects of the breakdown of the open structure and the increased thermal agitation produced by heating the system change predominance at a temperature of 3·98°C at which point water has its maximum density. Above this temperature the thermal expansion outbalances the effects of hydrogen bond collapse.

10.3 The Nature of the Hydrogen Bond

The strength of a hydrogen bond is intermediate between the Van der Waals cohesion and that of chemical bonds. The exact nature of the linkage is not clear but it may be understood in terms of several contributing effects. First there is the Van der Waals type dipole–dipole interaction between two adjacent molecules which, in the case of the liquid water system, may be written as shown in Figure 10.4, in which two water

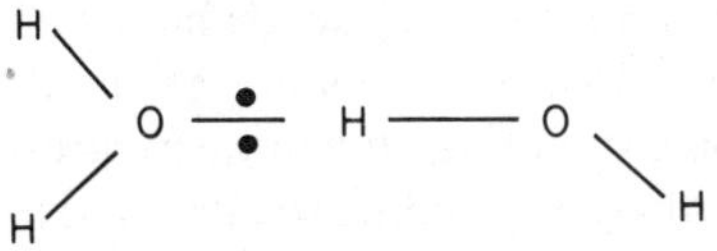

The Solid State and the Bonding in Ionic Compounds and Metals

Figure 10.4. Possible arrangement of two hydrogen bonded water molecules

molecules are arranged with one of the non-bonding pairs of electrons on an oxygen atom of one molecule directed towards a hydrogen atom of the other molecule.

Given this situation we may now consider the bridging hydrogen atom or, in particular, its nucleus. This single proton is able to exert a great coulombic effect upon the oxygen atom (in particular its lone pair) of the adjacent molecule. This is because the proton is taking part only in a polarized bond in which the maximum electron density is mainly localized near the oxygen atom of that bonding system. The bonding electrons, in other words, exert very little shielding of the proton as far as any adjacent molecules are concerned. The almost bare proton coupled with electronegative 'other atoms' seems to be the ingredients of a hydrogen bonding system.

The linear but asymmetrical hydrogen bonds seem therefore to be an extreme case of dipole–dipole interaction.

There are non-linear hydrogen bonds to which the same explanation applies and two examples will be quoted. These are the hydrogen bonds in *ortho*-salicylaldehyde and *ortho*-nitrophenol (I and II), as shown in Figure 10.5. These have melting points which are $-7°C$ and $45°C$ respectively.

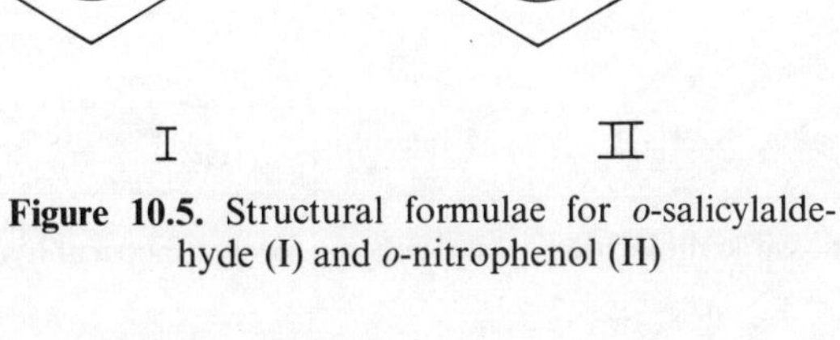

Figure 10.5. Structural formulae for *o*-salicylaldehyde (I) and *o*-nitrophenol (II)

The isomeric *meta* and *para* compounds have melting points shown in Table 10.3. The *meta* and *para* compounds are unable to take part in the *intramolecular* hydrogen-bonding shown in Figure 10.5, but may undertake *intermolecular* hydrogen-bonding to give solids with greater cohesion and, in consequence, ones with higher melting points.

Table 10.3. Melting Point Data for Solids which are Intermolecularly Hydrogen Bonded

Compound	m.p. (°C)
m-hydroxybenzaldehyde	106
p-hydroxybenzaldehyde	116
m-nitrophenol	96
p-nitrophenol	114

The linear symmetrical hydrogen bond in the ion, HF_2^-, is different from the other systems in its strength. The bond strength is 30 kcal.mole^{-1} (126 kJ.mole^{-1}). In other words to produce a fluoride ion and a neutral hydrogen fluoride molecule from an HF_2^- ion requires such an amount of energy. This is considerably greater than the energies of the asymmetric hydrogen bonds (3–7 kcal.mole^{-1}, 13–29 kJ.mole^{-1}), and may be interpreted in terms of a molecular orbital system. If we consider the fluorine atoms to contribute one p orbital (or an sp hybrid orbital) each and the 1s orbital of the hydrogen atom is used, we find the resulting molecular orbitals to be as shown in Figure 10.6. A linear three-centre system is produced in which (in HF_2^-) there are four electrons to be accommo-

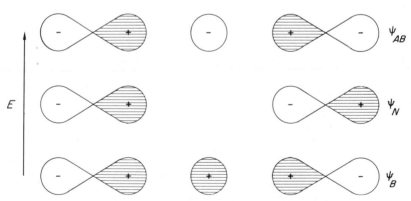

Figure 10.6. A possible molecular orbital scheme for symmetrical hydrogen bonds

The Solid State and the Bonding in Ionic Compounds and Metals

dated. These will normally have the configuration, $\psi_B^2 \psi_N^2$, thus leading to individual F—H bond orders of $\frac{1}{2}$.

As was the case with the xenon fluorides in which a similar type of bonding was proposed, it is essential that the atom other than hydrogen (or xenon) should be very electronegative. This is because in the systems,

$$F \cdots\cdots H^- \cdots\cdots F$$

and

$$F \cdots\cdots Xe \cdots\cdots F$$

electron density has to be transferred partially to the fluorine atoms (since two electrons occupy ψ_N which has no contribution from the central atom). Such transfer is most likely with fluorine (the most electronegative element). The ionization potentials of H^- and Xe are 0·78 ev and 12·13 ev respectively. Helium, which is isoelectronic with the hydride ion, has a rather large first ionization potential (24·58 ev) and would not therefore be expected to take part in such bonding. HeF_2 does not exist.

10.4 Covalent Solids

The principles concerning the bonding and shapes of simple polyatomic molecules (Chapters 8 and 9) may be applied to systems of greater complexity by considering them in terms of their basic units. Of many such systems only two will be considered in detail, these being elemental carbon in its diamond and graphite allotropic modifications.

Diamond

In order to explain the tetrahedral distribution of the four equivalent bonds in methane, it is necessary to introduce the concept of the hybridization of the atomic orbitals in the valency shell, sp^3 hybrid orbitals then being used by the carbon atom. The bonding of the carbon atoms in diamond is extremely similar to that in methane except that the hydrogen

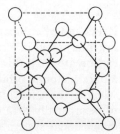

Figure 10.7. The diamond structure

atoms are all replaced by 'tetrahedral' carbon atoms and the sigma bonds in the system are sp^3–sp^3 type instead of sp^3–s as in methane.

The crystal structure of the diamond lattice is shown in Figure 10.7. In diamond the bond angles are exactly tetrahedral (109° 28′) and, since the whole crystal is held together by strong chemical (covalent) forces the high melting point of diamond may be understood.

Graphite

In graphite, the alternative allotropic modification of elemental carbon, the carbon atoms are arranged very differently. Graphite possesses a layer structure in which there is a gap of 0·35 nm between the planar layers. In the individual layers the carbon atoms are arranged hexagonally (as in benzene), as shown in Figure 10.8. The bond angles are 120° within the

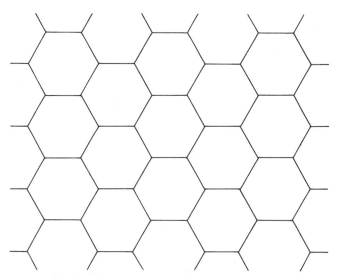

Figure 10.8. The hexagonal arrangement of atoms in a graphite layer

layers and by comparison with the bonding in benzene (Section 9.7) these are interpreted in terms of sp^2 hybridization of the atomic orbitals of the carbon atoms. If this is the case, then there remains on each carbon atom a pure p atomic orbital which does not contribute to the sigma-bonding scheme. These p orbitals are perpendicular to the two-dimensional sheets of hexagonally arranged carbon atoms and may participate in a two-dimensional system of delocalized pi molecular orbitals (cf. butadiene,

The Solid State and the Bonding in Ionic Compounds and Metals

etc., in Section 9.7). The extensive delocalization of the pi electrons in these orbitals is sufficient to stabilize the structure without involving any chemical interaction to occur between adjacent layers. The distance between adjacent layers would seem to rule out any covalent link and therefore it may be assumed that Van der Waals forces are responsible for such cohesion as there is.

The distance between adjacent carbon atoms within the layers is 0·142 nm (0·139 nm in benzene) which corresponds to a C—C (inplane) bond order of 1·33. The sigma bonding electrons are responsible for a bond order of 1·0 and the pi electrons which exist in the delocalized molecular orbital system have a range of bonding effects which contribute 0·33 to the overall bond order. The existence of electrons in the delocalized molecular orbital system is responsible for the electrical conduction properties of graphite. A perfect crystal of graphite is an insulator in a direction perpendicular to the sheets, but a conductor across the sheets. Both forms of elemental carbon may be described as giant molecules, there being no individually distinguishable discrete smaller units than the whole crystal which is therefore the molecule of carbon. Such considerations of course apply to ionic and covalent crystals in which there are no discrete 'molecules'.

10.5 Ionic Solids

When two atoms combine together to form a single bond it is usual for two electrons to occupy a bonding molecular orbital (Section 7.3). If the two atoms concerned in the bonding are not identical, then it is probable that the energies of the two atomic orbitals are not equal and that one (the lower in energy) will contribute a major fraction to the bonding molecular orbital (Section 7.8). If the difference in energy between the two orbitals is very large, then the bonding molecular orbital will be almost identical to the atomic orbital of lower energy and, if the two electrons occupy this level, it indicates the transfer of an electron from one atom to the other resulting in the production of an ionic compound. The influence of the energy difference upon the type of bond produced is summarized by the diagrams in Figure 10.9.

Figure 10.9(a) shows the situation for the production of a homopolar diatomic molecule (e.g. H_2) in which ψ_1 and ψ_2 contribute equally to ψ_B and ψ_{AB}:

$$\psi_B = \psi_1 + \psi_2 \qquad (10.2)$$

and

$$\psi_{AB} = \psi_1 - \psi_2 \qquad (10.3)$$

Atomic and Molecular Structure

If there is a difference in energy between the two contributing orbitals as in Figure 10.9(b), then the lower orbital, ψ_2, will contribute a greater

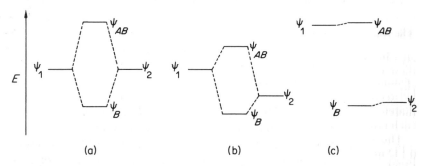

Figure 10.9. Representation of the formation of bonding and antibonding orbitals when two atomic orbitals interact which (a) are identical in energy, (b) are different in energy, and (c) are very different in energy

amount to the bonding orbital, ψ_B. The molecular wave functions would be

$$\psi_{AB} = \lambda\psi_1 - \psi_2 \tag{10.4}$$

$$\psi_B = \psi_1 + \lambda\psi_2 \tag{10.5}$$

with the coefficient having a value such that $\lambda^2 > 1$.

A situation like this is encountered where the two elements concerned differ in electronegativity (Section 7.7) as in HF, and the covalent bond produced may be said to be polarized, indicating an asymmetric distribution of electronic charge. Such molecules are dipolar and have a non-zero dipole moment. In Figure 10.9(c) the situation is taken to its extreme limit where ψ_B and ψ_2 are identical and ψ_1 and ψ_2 differ by an amount of energy which is too large to allow their mixing. In this case electron transfer from atom A to atom B occurs which results in the formation of ions.

The formation of an ionic compound is more complex than the treatment so far indicates. The reason for this is that, although in the cases shown in Figure 10.9(a) and (b) the molecules produced (i.e. H_2 and HF) do exist as diatomic molecules in the case of the ionic compound, ion pairs A^+B^- *do not* normally have a separate stable existence. Ionic compounds are found to be solid crystalline substances in which the

The Solid State and the Bonding in Ionic Compounds and Metals

atoms or ions are arranged in a regular three-dimensional array which is called the crystal lattice. It is not possible to distinguish individual formula units.

Results of X-ray diffraction experiments with sodium chloride crystals show that electron densities in the region of the sodium and chlorine nuclei are consistent with complete electron transfer from sodium to chlorine and that Na^+ and Cl^- ions exist. Also the electron density in the internuclear regions is found to be low, which is not the case for covalent bonding. In the liquid state (molten) sodium chloride conducts electricity. X-ray crystallographic investigations of the sodium chloride crystal have shown the arrangement of ions to be a cubic system in which there are alternately sodium and chloride ions, each ion having a co-ordination number (number of nearest neighbours) of six. The simplest representation of the structure which shows the maximum symmetry is shown in Figure 10.10.

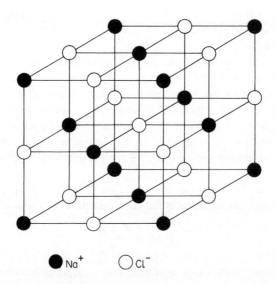

Figure 10.10. The usual representation of the sodium chloride lattice Na^+ (●) Cl^- (○)

The detailed treatment of the feasibility of using an ionic model to describe compounds such as sodium chloride has two aspects—experimental and theoretical.

10.6 Theoretical Treatment of Ionic Bonding

If we consider the case of sodium chloride as being one involving two types of ion—one having an electronic charge, $+Z_1e$ (where $Z_1 = 1$), and the other having an electronic charge, $-Z_2e$ (where $Z_2 = 1$)—and as a compound having the three-dimensional lattice shown in Figure 10.10, we may calculate a value for the *lattice energy* in terms of the attractive and repulsive forces operating between the ions involved. The lattice energy of a crystal may be defined as the amount of energy *released* when the gaseous ions condense to form the crystal.

The *attractive* force is the coulombic one which operates between ions of opposite charge. There is also a coulombic *repulsive* force in operation between the ions of like charge as well as a *short range* repulsive force operating as a result of the proximity of the filled orbitals of the ions. Both ions have inert-gas electronic configurations (Na^+, $1s^2\ 2s^2\ 2p^6$ and Cl^-, $1s^2\ 2s^2\ 2p^6\ 3s^2\ 3p^6$) in which all the low lying atomic orbitals are filled, and in bringing together two such ions a point must be reached when orbitals (filled) begin to overlap. This is a clear contravention of the Pauli Exclusion Principle (Sections 6.2 and 3) and is the reason for the repulsive force.

Other forces operating are the 'intermolecular' or Van der Waals forces (Section 10.2) which contribute in a minor way to the cohesion. These, like the repulsive force, operate only at short range.

When the gaseous ions, $NA^+_{(g)}$ and $Cl^-_{(g)}$, are brought from infinite separation (which we take as the arbitrary zero of energy) the potential energy of the system is lowered by the attractive forces and increased by the repulsive forces (cf. Figure 7.8 for the production of the hydrogen molecule).

The mathematical form of the variation of potential energy with internuclear (interionic) distance is for the coulombic attraction, U, given by

$$U = -\frac{Z_1 Z_2 e^2}{r} \quad (10.6)$$

for two particles of charges $+Z_1e$ and $-Z_2e$ approaching from an infinite separation to the distance apart, r. The repulsive force is given by

$$U_{rep} = \frac{B}{r^n} \quad (10.7)$$

where the exponent, n, is derived from measurements of the compressibility of the crystal and has the value 9 for sodium chloride and where B is an

The Solid State and the Bonding in Ionic Compounds and Metals

undetermined constant. To use equation (10.6) to represent the total coulombic force contribution would be incorrect since it gives the lowering of potential energy for an ion pair. In the crystal lattice of sodium chloride each sodium ion possesses six nearest neighbours of opposite charge at a distance, r, and twelve next nearest neighbours of the same charge at a distance, $\sqrt{2}r$, as well as eight chloride ions at a distance, $\sqrt{3}r$, and six sodium ions at a distance, $2r$, and twenty-four chloride ions at a distance, $\sqrt{5}r$, etc., where r represents the minimum distance between adjacent sodium and chloride ions.

The total coulombic contribution (attractions and repulsions) to the potential energy is thus given by

$$U = -\frac{6Z_1Z_2 e^2}{r} + \frac{12Z_1Z_2 e^2}{\sqrt{2}r} - \frac{8Z_1Z_2 e^2}{\sqrt{3}r} + \frac{6Z_1Z_2 e^2}{2r}$$

$$-\frac{24Z_1Z_2 e^2}{\sqrt{5}r} + \cdots \tag{10.8}$$

which may be written as:

$$U_{\text{coulombic}} = -\frac{Z_1Z_2 e^2}{r}\left(6 - \frac{12}{\sqrt{2}} + \frac{8}{\sqrt{3}} - \frac{6}{2} + \frac{24}{\sqrt{5}} - \cdots\right) \tag{10.9}$$

The series in parentheses may be extended to cover the whole crystal and has a value of 1·74756 for sodium chloride. It would of course have other values for other stereochemical arrangements. This number is known as the Madelung Constant (or Factor) for this particular crystal lattice. Equation (10.8) may be rewritten as:

$$U_{\text{coulombic}} = -\frac{Z_1Z_2 e^2 M}{r} \tag{10.10}$$

where M is the Madelung Constant.

The total potential energy is given by

$$U = U_{\text{coulombic}} + U_{\text{repulsive}}$$

$$= -\frac{Z_1Z_2 e^2 M}{r} + \frac{B}{r^n} \tag{10.11}$$

As a result of the attractive and repulsive contributions to the potential energy there is a certain interionic distance, r_{eq}, where the system has minimum potential energy. The distance, r_{eq}, represents the equilibrium interionic separation in the crystal lattice. At the minimum of the potential energy curve (Figure 7.8) where $r = r_{eq}$ we may write:

$$\left(\frac{dU}{dr}\right)_{r=r_{eq}} = 0 \qquad (10.12)$$

and so we may differentiate equation (10.11) and obtain:

$$\frac{dU}{dr} = \frac{Z_1 Z_2 e^2 M}{r^2} - \frac{nB}{r^{n+1}} \qquad (10.13)$$

and putting $r = r_{eq}$ this becomes:

$$\frac{Z_1 Z_2 e^2 M}{r_{eq}^2} = \frac{nB}{r_{eq}^{n+1}} \qquad (10.14)$$

which can be rearranged to give

$$B = \frac{Z_1 Z_2 e^2 M}{n} r_{eq}^{n-1} \qquad (10.15)$$

so that equation (10.11) becomes

$$U = \frac{Z_1 Z_2 e^2 M}{r} + \frac{Z_1 Z_2 e^2 M r_{eq}^{n-1}}{n r^n} \qquad (10.16)$$

which may be simplified to read:

$$U = \frac{Z_1 Z_2 e^2 M}{r}\left[1 - \frac{1}{n}\left(\frac{r_{eq}}{r}\right)^{n-1}\right] \qquad (10.17)$$

The lattice energy which we may define as the energy released in the reaction,

$$Na^+_{(g)} + Cl^-_{(g)} \rightarrow Na^+Cl^-_{cryst} \qquad (10.18)$$

i.e. the heat of formation of the solid crystal from its gaseous ions will be equal to $-U_{total}$, and expressed in terms of the gram-mole becomes

$$L = -U_{total}N = +\frac{Z_1 Z_2 e^2 MN}{r_{eq}}\left[1 - \frac{1}{n}\right] \qquad (10.19)$$

since $r = r_{eq}$ for the crystal in its equilibrium state. N is Avogadro's number and $n = 9$ for sodium chloride. Equation (10.19) is known as the Born–Landé equation and was modified by Born and Mayer who, instead of using equation (10.7) for the repulsion term, substituted the equation,

$$U_{repulsion} = be^{-r/\rho} \qquad (10.20)$$

The Solid State and the Bonding in Ionic Compounds and Metals

There is some wave-mechanical justification for this since the radial functions of atoms have a similar mathematical form. The use of this function in which b and ρ are constants produces the Born–Mayer expression for the lattice energy of a crystal as

$$L = -U_{\text{total}}N = +\frac{Z_1 Z_2 e^2 MN}{r_{\text{eq}}}\left[1 - \frac{\rho}{r_{\text{eq}}}\right] \qquad (10.21)$$

In this equation the constant ρ is taken as 0·0311 nm.

For the calculation of the lattice energy of sodium chloride we have $Z_1 = 1, Z_2 = 1, e = 4{\cdot}803 \times 10^{-10}$ e.s.u., $M = 1{\cdot}74756, N = 6{\cdot}023 \times 10^{23}$, $r_{\text{eq}} = 0{\cdot}2819 \times 10^{-9}$ m and $\rho = 0{\cdot}0311 \times 10^{-9}$ m. These values substituted into equation (10.21) give a value of 183 kcal.mole^{-1} (767 kJ.mole^{-1}) for the lattice energy of sodium chloride.

In this relatively straightforward calculation of the lattice energy of sodium chloride the small contribution to cohesion from intermolecular forces has been ignored. That there is a contribution is not in doubt and it would be a London force contribution, since the ions concerned are spherically symmetrical.

A quantity such as the lattice energy is necessarily difficult to estimate by direct observation because of the difficulty in producing a system which involves equilibrium between the gaseous atoms and ions concerned. There is, however, an indirect experimental method of estimating lattice energies which also gives insight into the factors which influence ionic bond formation.

10.7 'Experimental' Treatment of Ionic Bonding: Born–Haber Thermochemical Cycles

When one gram-mole of sodium chloride is produced in its standard state, i.e. crystalline solid (indicated by the subscript, c), by the reaction,

$$\text{Na}_{(s)} + \tfrac{1}{2}\text{Cl}_{2(g)} \rightarrow \text{Na}^+\text{Cl}^-_{(c)} \qquad (10.22)$$

the change in heat content is $\Delta H = -98{\cdot}2$ kcal.mole^{-1} (-411 kJ.mole^{-1}). The negative sign implies that the reaction is exothermic and that the energy is released when the sodium chloride is formed. Since the reactants in equation (10.22) are elements in their respective standard states, the change in heat content may be regarded as the heat of formation of sodium chloride.

The overall reaction, equation (10.22), may be split up into several stages for which a ΔH value is known. These stages are:

(1) the sublimation of the solid sodium:

$$Na_{(s)} \to Na_{(g)},$$

$$\Delta H_S = +26{\cdot}0 \text{ kcal.mole}^{-1} (+109 \text{ kJ.mole}^{-1}) \quad (10.23)$$

(2) the dissociation of the molecular chlorine gas:

$$\tfrac{1}{2}Cl_{2(g)} \to Cl_{(g)},$$

$$\Delta H_D = \tfrac{1}{2}D = +29{\cdot}0 \text{ kcal.mole}^{-1} (+121{\cdot}5 \text{ kJ.mole}^{-1}) \quad (10.24)$$

(D is the dissociation energy of the chlorine molecule.)

(3) the ionization of the gaseous sodium atoms:

$$Na_{(g)} \to Na^{+}{}_{(g)} + e^{-},$$

$$\Delta H_I = I_{Na} = +118 \text{ kcal.mole}^{-1} (+494 \text{ kJ.mole}^{-1}) \quad (10.25)$$

(I_{Na} is the ionization potential of sodium.)

(4) the production of chloride ions:

$$e^{-} + Cl_{(g)} \to Cl^{-}{}_{(g)},$$

$$\Delta H_E = -E_{Cl} = -86{\cdot}5 \text{ kcal.mole}^{-1} (-362 \text{ kJ.mole}^{-1}) \quad (10.26)$$

(E_{Cl} is the electron affinity of chlorine and this amount of energy is released.)

(5) the aggregation of the gaseous ions to form the crystal lattice:

$$Na^{+}{}_{(g)} + Cl^{-}{}_{(g)} \to Na^{+}Cl^{-}{}_{(c)}, \quad \Delta H_L = -L \quad (10.27)$$

(L is the lattice energy released in this reaction.)

Applying the First Law of Thermodynamics to the system, we may equate the sum of the heat content changes in steps (1) to (5) with the heat of formation ΔH_f of $-98{\cdot}2$ kcal.mole^{-1} (-411 kJ.mole^{-1}):

$$-98{\cdot}2 = \Delta H_S + \Delta H_D + \Delta H_I + \Delta H_E + \Delta H_L \quad (10.28)$$

and putting in the values for these we get

$$-98{\cdot}2 = 26{\cdot}0 + 29{\cdot}0 + 118{\cdot}0 - 86{\cdot}5 - L \quad (10.29)$$

from which we calculate the value of L to be 184·7 kcal.mole^{-1} (772 kJ.mole^{-1}), which is in very reasonable agreement with the Born–Mayer calculated value.

The formation reaction, equation (10.22), and the five stages into which it may be split may be represented in the form of a Born–Haber thermochemical cycle, viz.:

The Solid State and the Bonding in Ionic Compounds and Metals 257

$$
\begin{array}{ccccc}
Na_{(s)} & + & \tfrac{1}{2}Cl_{2(g)} & \xleftarrow{\Delta H = +98\cdot 2} & Na^{+}Cl^{-}_{(c)} \\
| & & | & & \uparrow \\
\Delta H_S = +26\cdot 0 & & \Delta H_D = +29\cdot 0 & & \Delta H = -L \\
\downarrow & & \downarrow & & | \\
Na_{(g)} & + & Cl_{(g)} & \xrightarrow{\Delta H = 118 - 86\cdot 5} & Na^{+}_{(g)} + Cl^{-}_{(g)}
\end{array}
$$

Independently of the starting point in such a cycle, the overall sum of the changes in heat content must be zero which leads to the equation,

$$98\cdot 2 + 26\cdot 0 + 29\cdot 0 + 118 - 86\cdot 5 - L = 0 \qquad (10.30)$$

which upon rearrangement gives equation (10.29) and, of course, the same value for L.

The purpose of applying this Born–Haber cycle treatment to the formation of sodium chloride is to show that, within the limits of the experimental measurements made, the theoretical idea of the compound being composed of an array of positive and negative ions is well founded. In fact, the thermochemical measurements in addition to the calculation of lattice energy were used initially for the determination of electron affinities. The great value of the cycle is that it allows the statement of the factors which favour the formation of ionic compounds. These are that:

(1) the sublimation energy of the metal should be low,
(2) the dissociation energy of the element giving rise to the negative ion should be small,
(3) the ionization potential of the metal should be low,
(4) the electron affinity of the element forming the negative ions should be high, and
(5) the lattice energy should be large.

These conditions are those which maximize the exothermicity of the formation reaction. By using the Born–Haber cycle treatment it is possible to predict the heat of formation of a compound. If the prediction agrees with the observed value, then the compound may be regarded as being ionic. If the compound is other than ionic, the predicted heat of formation will be less negative and more positive than the observed value. The argument for the preceding statement is that a system normally will exist in its lowest energy state. If this state is other than ionic then it is obvious that an ionic description will not be satisfactory. It would not be expected that the ΔH_f predicted for an ionic substance would agree with the experimental ΔH_f value. This may be summarized as in the diagram of Figure 10.11. In Figure 10.11(a) it is shown that ΔH_f as predicted by a Born–Haber cycle using a calculated lattice energy (Born–Mayer) is appreciably less than the actual value. The compound to which these values apply may not be taken to be a completely ionic one whereas, if the situation as shown in Figure 10.11(b) exists, then the prediction of ionicity is valid.

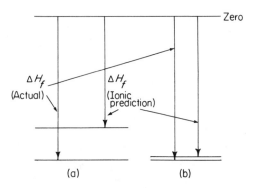

Figure 10.11. Comparisons of ΔH_f (actual) and ΔH_f (predicted for ionic compound) for two situations, (a) where the ionic description is not good, and (b) where the ionic description is almost perfect

Approximate calculations may be carried out for the chlorides of magnesium and aluminium, assuming them to be completly ionic.

Magnesium Chloride

An additional point may be dealt with at this stage, which is the factors governing the stoichiometry of the compound formed. With covalent bond formation the composition of compounds depends on the number of unpaired electrons on the central atom, e.g. in the oxygen ground state there are two unpaired electrons and the stoichiometry of the hydride is OH_2. In ion formation we are not necessarily concerned with electron pairing as the only factor influencing the stoichiometry and we have to consider the feasibility of such compounds as $NaCl_2$ (involving the Na^{2+} ion) and $MgCl$ (involving the Mg^+ ion). The removal of a second electron from sodium to form the doubly charged cation requires the expenditure of $1090 \text{ kcal.mole}^{-1}$ ($4560 \text{ kJ.mole}^{-1}$) which would vastly outweigh the factor of two increase in lattice energy (all other factors being equal). A compound such as $NaCl_2$ is therefore not feasible thermodynamically since its ΔH_f would be very positive. It would be given by

$$\Delta H_f(NaCl_2) = \Delta H_S + \Delta H_D + I_1 + I_2 - 2E_{Cl} - L_{NaCl_2}$$
$$= 26 + 58 + 118 + 1090 - (86 \cdot 5 \times 2) - (2 \times 184)$$
$$= +751 \text{ kcal.mole}^{-1} \ (3150 \text{ kJ.mole}^{-1}) \quad (10.31)$$

The compound, if produced, would be unstable with respect to the disproportionation reaction,

$$NaCl_2 \rightarrow NaCl + \tfrac{1}{2}Cl_2 \quad (10.32)$$

which would have a ΔH value of -849 kcal.mole^{-1} (-3550 kJ.mole^{-1}) (of NaCl produced).

In the case of MgCl we may calculate the heat of formation for a compound with a sodium chloride structure ($M = 1.75$) and with an interionic distance of 0·28 nm. This would be given by

$$\Delta H_f(MgCl) = \Delta H_S + \Delta H_D + I_{Mg} - E_{Cl} - L_{MgCl} \quad (10.33)$$

The lattice energy would be about 184 kcal.mole^{-1} (770 kJ.mole^{-1}), and this, together with the other quantities, would give

$$\Delta H_f(MgCl) = 35 \cdot 6 + 29 \cdot 0 + 176 \cdot 0 - 86 \cdot 5 - 184$$
$$= -30 \text{ kcal.mole}^{-1} \ (-125 \cdot 5 \text{ kJ.mole}^{-1}) \quad (10.34)$$

which is a reasonably negative value and indicates the stability of such a compound compared to its constituent elements in their standard states.

Now consider the actual compound formed when magnesium reacts with chlorine, which has the formula $MgCl_2$. This has what is known as a Cadmium Iodide Structure (a layer lattice) for which the Madelung Constant is 2·35. The interionic distance is 0·246 nm from which we may calculate the lattice energy to be 567 kcal.mole^{-1} (2370 kJ.mole^{-1}). The second ionization potential of magnesium is 346 kcal.mole^{-1} (1450 kJ.mole^{-1}) and the heat of formation of ionic $MgCl_2$ is calculated to be

$$\Delta H_f(MgCl_2) = \Delta H_S + \Delta H_D + I_1 + I_2 - 2E_{Cl} - L$$
$$\Delta H_f(MgCl_2) = 35 \cdot 6 + 58 \cdot 0 + 176 + 346 - (2 \times 86 \cdot 5) - 567$$
$$= -114 \cdot 4 \text{ kcal.mole}^{-1} \ (-479 \text{ kJ.mole}^{-1}) \quad (10.35)$$

The data for ionic MgCl, ionic $MgCl_2$ and actual $MgCl_2$ are shown in the diagram in Figure 10.12. The actual heat of formation of magnesium chloride is $-153 \cdot 4$ kcal.mole^{-1} (-640 kJ.mole^{-1}) and indicates that the

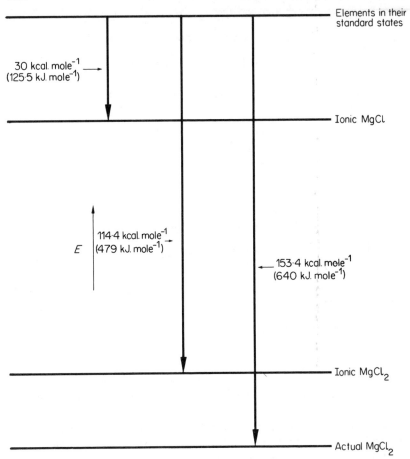

Figure 10.12. A comparison of the heats of formation of ionic MgCl and ionic $MgCl_2$ with that of actual $MgCl_2$, the reference being the standard states of the elements

ionic description is not wholly correct. The compound Mg^+Cl^- would certainly disproportionate according to the equation,

$$2MgCl_{(c)} \rightarrow MgCl_{2(actual)} + Mg_{(s)} \tag{10.36}$$

which reaction would have a ΔH of -213 kcal.mole^{-1} (-890 kJ.mole^{-1}).

To summarize the factors most influencing the stoichiometry of an ionic compound we have, in the case of sodium chloride, shown that

The Solid State and the Bonding in Ionic Compounds and Metals

increased ionization produces instability while in the case of magnesium chloride the reverse is the case. However, with $MgCl_3$ involving Mg^{3+} ions again the increase in ionization energy confers instability on the product. The dividing line between the amount of ionization which occurs in actual compounds and that which confers instability is intimately concerned with the electronic configuration of the elements involved. The successive ionization potentials of some of the elements of the second short period are shown in Table 10.4. The stepwise divisions shown in

Table 10.4. The Successive Ionization Potentials of the Elements, Na to P

Element	Ionization potential (kcal.mole^{-1})					
	1	2	3	4	5	6
Na	118	1090	1660	2280	3200	3970
Mg	176	346	1850	2520	3260	4310
Al	138	434	658	2770	3540	4390
Si	188	377	772	1042	3830	4700
P	242	455	697	1185	1500	5150

Table 10.4 separate the ionizations (to the left) which are from the $n = 3$ levels from those (to the right) which are from the much more stable $n = 2$ levels. The ionizations to the left are those which could possibly be expected to occur in ionic bond formation. Whether actual ionic bond formation occurs depends largely on whether the total ionization energy is smaller than the expected lattice energy of the compound. Even for the Al^{3+} ion forming an ionic chloride this is not so, and the predicted heat of formation of $AlCl_3$ (ionic) is $+272$ kcal.mole^{-1} (1140 kJ.mole^{-1}), whereas actual aluminium chloride is a covalently bound dimeric molecule, Al_2Cl_6, and has a heat of formation of -166 kcal.mole^{-1} (-694 kJ.mole^{-1}). In general, the greater the difference between ΔH_f (calculated) from ΔH_f (actual), the greater is the deviation of the bonding from 100 per cent ionic character.

In view of the above discussion the factors presented as favouring the production of ionic compounds need a slight revision. The fifth condition, that the lattice energy should be large, must be modified since a large lattice energy is only attained by having high ionic charge which, as we have seen, leads to covalency. The ionization energy required to produce high charges is greater than the consequent increase in the lattice energy. The lattice energy should be as large as is consistent with a reasonably small ionization energy. The factors which lead to a greater lattice energy (i.e. higher charged cations and a small cation radius) are precisely those which also increase the ionization energy required to produce the compound. Since the ionization energy increases with successive ionizations by more than a linear rate, there is produced a situation where ionic bond formation is not energetically possible.

A note at this stage is required concerning the predictions of reactions based upon ΔH values. Such predictions are not thermodynamically correct since the parameter which does have predictive value is the change in the *free energy*, ΔG, which is defined by

$$\Delta G = \Delta H - T\Delta S \tag{10.37}$$

where T is the temperature of the system and ΔS represents the entropy change involved in the reaction. This latter term has been ignored in making the above predictions. It is thermodynamically accurate to say that a reaction is feasible if the change in free energy is negative, i.e. free energy is released. A negative ΔH value would, of course, contribute towards a negative ΔG value, and we may say that if a reaction possesses a very negative ΔH value there is a good possibility that ΔG also will be negative. It is, of course, possible for a reaction to have a positive value of ΔH and still be feasible (i.e. have a negative ΔG value) if $T\Delta S > \Delta H$. Such reactions are favoured by high temperatures.

The free energy, heat content, $T\Delta S$, and entropy changes involved in the formation of sodium chloride, magnesium chloride and aluminium chloride are shown in Table 10.5. The $T\Delta S_f$ term is of the order of

Table 10.5. Thermodynamic Data for Some Compounds

Compound	ΔH_f		$\Delta G_f{}^a$		$T\Delta S_f$	
	(kcal.-mole^{-1})	(kJ.-mole^{-1})	(kcal.-mole^{-1})	(kJ.-mole^{-1})	(kcal.-mole^{-1})	(kJ.-mole^{-1})
NaCl	−98·6	−413	−92·2	−386	−6·4	−27
MgCl$_2$	−153·4	−640	−141·6	−590	−11·8	−50
Al$_2$Cl$_6$	−166·2	−696	−148·2	−621	−18·0	−75

^{a} Values of $T\Delta S_f$ are for a temperature of 298°K.

6–11 per cent of the ΔH_f value, and the qualitative predictions of reactions based upon ΔH_f are justified in the above cases. It would be a very different matter if $T\Delta S_f$ were comparable in value to ΔH_f, in which case ΔG_f would be the only valuable guide to the feasibility of reaction.

The Solid State and the Bonding in Ionic Compounds and Metals

10.8 The Metallic Bond

Metals possess structures in which each atom is surrounded by either twelve or fourteen nearest neighbours. In other words the coordination numbers for metal atoms are high. The closeness of packing of atoms in the metallic state required to achieve such high coordination numbers is responsible for the high specific gravities of metallic elements. Other features of the metallic state which we may consider in this section are the general property of electrical conduction and the low values of the work function (Section 1.4).

With such high coordination numbers it is clear that the use of localized electron pair bonds is precluded since the metallic elements in general possess few valency electrons. In lithium metal each atom has a single valency electron, $2s^1$, the $1s^2$ pair being too stable for use in bonding. It is therefore only possible for lithium to participate in one electron pair bond as in the molecule Li_2 which exists in the gaseous state. In the solid state a different treatment has to be used. Molecular orbital theory can be applied to the metallic state. This makes use of the valency electrons of the metal atoms in forming molecular orbitals which are as extensive as the metal crystals themselves. In other words, the molecular orbitals are giant orbitals which extend throughout each individual crystal of the metal.

If two atoms combine, using one orbital from each, then two molecular orbitals are produced, one being a bonding orbital of low energy and the other being an antibonding orbital of high energy. If an array of N atoms combine, again using one orbital from each atom, then N molecular orbitals may be produced. One-half of these orbitals will have an excess of bonding character, the other half having an excess of antibonding character. Thus from N degenerate atomic orbitals there will be produced a band of N non-degenerate molecular orbitals. This is shown diagrammatically in Figure 10.13. The electrons in such a band of molecular orbitals will occupy the lowest available levels and will pair up in such levels.

The lithium atom, for example, has three electrons, $1s^2 2s^1$, and only the $2s$ electron can be considered for valency purposes since the $1s^2$ pair is very stable. The interaction of the $2s$ orbitals of a three-dimensional array of lithium atoms will produce a $2s$ band of molecular orbitals. One-half of the levels in this band will be filled at absolute zero. At normal temperatures there will be some of the higher levels singly filled. When an electrical potential gradient is applied to the lithium crystal it is possible for the electrons to be mobile in the $2s$ 'conduction band' and

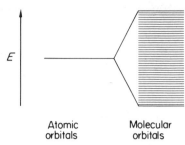

Figure 10.13. The formation of a bond of molecular orbitals from an array of identical atomic orbitals

to use singly occupied levels to transfer charge through the crystal. The elements of Group I of the periodic table are among the best conductors available.

A metal with two valency electrons, such as beryllium, with an outer electronic configuration, $2s^2$, might be expected to be an insulator since there are sufficient electrons to fill the $2s$ conduction band. It is true that Group II metals are not as good conductors of electricity as those of Group I, but nevertheless they are reasonably good conductors. To explain this it is necessary to consider the other orbitals of the same

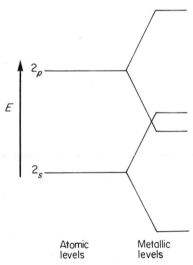

Figure 10.14. The overlapping of the energies of orbitals in the $2s$ and $2p$ bands of a metal

The Solid State and the Bonding in Ionic Compounds and Metals

principal quantum number value. There are the normally vacant $2p$ orbitals to consider. If these also formed a three-dimensional band this could very well overlap with the $2s$ band as shown in Figure 10.14.

If this is so then the valency electrons could occupy both the $2s$ and $2p$ bands, neither of which would be filled, and conduction of electricity would be understandable.

Band theory may be applied to insulators, such as diamond, with equal success. Here it is found that the lowest vacant band does not overlap with the lower filled bands and no metallic conduction can take place. The situation is illustrated by Figure 10.15.

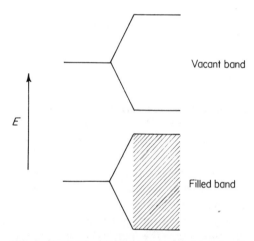

Figure 10.15. The highest energy filled band and the lowest energy vacant band of an insulator

The elements most likely to exist in the metallic state are those which have low values for their ionization potentials. Such elements are to be found at the left-hand side of the periodic table (Figure 6.10), since along any period the effective nuclear charge increases from left to right. That is, the ionization energy increases along each period (Figure 2.12, for example), and the electropositive elements are to be found at the beginning of each period.

In the groups of the periodic table, the effective nuclear charge decreases as each electron shell is filled and electropositive character increases down each group.

These electropositive elements, then, are the ones expected to be able to participate in extensive delocalization of their valency electrons and to exist, therefore, as metals. Due to their low ionization potential values there is an energetic advantage in such a bonding scheme.

There is not a well-defined boundary between metals and non-metals and certain elements possess neither complete metallic character nor complete non-metallic character. Such borderline elements include germanium, arsenic, antimony, tin, tellurium, polonium and astatine. Of these we may take tin as an example of where the metallic and non-metallic character of the material depend on the particular crystal structure of the allotropic modification. Tin normally exists in a metallic form, but at temperatures below 13·1° a transition is possible to grey tin which has the diamond arrangement (Figure 10.7) and which does not exhibit metallic conduction.

The metallic state, then, involves easily ionized atoms (with low work functions) which have their valency electrons delocalized in a molecular orbital band arrangement. This allows for high coordination numbers (and therefore high densities) since no discrete covalent bonds are required. It allows for electrical conduction in which the increasing vibration of the atoms interferes with electron flow as the temperature increases. It also is consistent with the ductility and maleability of metals in which adjacent layers slide easily with respect to each other and is evidence of fairly weak bonding. In a molecular orbital metallic band only the lower energy valency electrons are in very good bonding orbitals, and, on average, the metallic bond is weaker than normal covalent bonds occurring between two atoms.

It has already been pointed out that if the various molecular orbital bands do not overlap then the particular material is an insulator. Insulators have extremely low electrical conductivities. Between insulators and metallic conductors there are materials which can, under various circumstances, conduct electricity reasonably well. These materials are known as *semiconductors* and exhibit electrical conductivity which increases with increasing temperature.

In outline, the conduction processes in a semiconductor may be understood in terms of band theory. Various situations need to be considered.

(1) If the band energies are such that the energy gap is small, as in Figure 10.16(a) with the lower band completely full, an insulator would result. At temperatures sufficient to cause thermal excitation of electrons from the filled band to the upper, otherwise vacant, band a semiconductor is produced. Both bands may then be used to conduct current and the conductivity will increase as the temperature increases corresponding to a greater population of the

The Solid State and the Bonding in Ionic Compounds and Metals

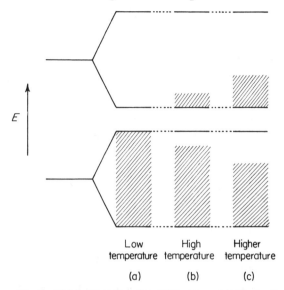

Figure 10.16. Showing the different populations of two bands of a semiconductor: (a) at low temperature where the material is an insulator, (b) at higher temperature where some thermal excitation of electrons has occurred, and (c) at an even higher temperature showing the greater population of the upper band

upper band. The low and high temperature cases are shown in Figure 10.16.

An alternative method of populating the upper band is to excite the electrons by using quanta of electromagnetic radiation. When an insulator shows semiconductor properties when illuminated, such behaviour is known as photoconductivity.

(2) Another way in which semiconduction properties may be achieved is if in an insulator there is an impurity present, some of whose electron energy levels lie in energy between a vacant and a filled band. There are two cases to be discussed, both being illustrated by Figure 10.17.

(a) If the impurity levels are originally occupied by electrons, then these electrons may be thermally excited to populate the originally vacant band of the insulator. Semiconductor properties result from this excitation, the conduction being carried out by the electrons accepted from the impurity levels. This kind of semiconduction is known as normal or n-type.

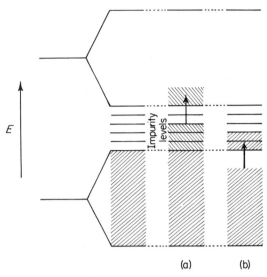

Figure 10.17. Showing the bridging of the filled and vacant bands of a semiconductor by impurity energy levels. Semiconduction can be achieved: (a) by filled impurity levels populating the originally vacant band, or (b) by population of vacant impurity levels by electrons originally occupying the filled band of the semiconductor

(b) If the impurity levels, which bridge the two bands of the insulator, are normally vacant, then they may accept electrons from the filled band of the insulator to give the situation in Figure 10.17. The upper band remains vacant. The transfer of electrons from the lower band to the impurity levels may be regarded as producing positive 'holes' in the lower band. The conduction which can then take place in the lower band is known as positive hole (p) type. One can either regard conduction as the movement of electrons in one direction or, as an alternative, the movement of positive holes in the opposite direction.

Problems

10.1 State the type of force responsible for the cohesion of the following solids: carbon dioxide, rubidium fluoride, polyethylene, tin, phenol, t-butanol, copper, anthracene, ammonium chloride and silicon carbide.

The Solid State and the Bonding in Ionic Compounds and Metals

10.2 Calculate the lattice energy of calcium fluoride, given that the equilibrium interionic distance is 0·236 nm and that the appropriate Madelung Constant is 2·51939.

10.3 Calculate the lattice energy of tin(IV) oxide, given that the equilibrium interionic distance is 0·218 nm and that the appropriate Madelung Constant is 2·408.

10.4 From the following data (in kcal.mole^{-1}) and the results of problems (10.2) and (10.3), calculate the heats of formation of the compounds, calcium fluoride and tin(IV) oxide.

Sublimation energies	calcium	46
	tin	72
Ionization energies	$Ca^0 \rightarrow Ca^{II}$	418
	$Sn^0 \rightarrow Sn^{IV}$	2155
Electron affinities	$F(0) \rightarrow F(-1)$	83
	$O(0) \rightarrow O(-2)$	155
Bond dissociation energies	$F_{2(g)} \rightarrow 2F_{(g)}$	38
	$O_{2(g)} \rightarrow 2O_{(g)}$	118

Compare your answers with the observed heats of formation of calcium fluoride ($-290\cdot3$) and tin(IV) oxide ($-138\cdot8$).

10.5 Plot the radii (given in nm) of the following ions against the atomic number of the atoms from which they are derived:

Li^+	0·068	Be^{2+}	0·03	F^-	0·133	O^{2-}	0·145	Al^{3+}	0·045
Na^+	0·098	Mg^{2+}	0·065	Cl^-	0·181	S^{2-}	0·190	Ga^{3+}	0·060
K^+	0·133	Ca^{2+}	0·094	Br^-	0·196	Se^{2-}	0·202	In^{3+}	0·081
Rb^+	0·148	Sr^{2+}	0·110	I^-	0·219	Te^{2-}	0·222	Tl^{3+}	0·091
Cs^+	0·167	Ba^{2+}	0·129						

Make correlations of the values corresponding to (a) isoelectronic ions, and (b) ions of the same charge. Discuss your correlations in terms of the effective nuclear charges of these ions (see Appendix I).

10.6 The electron affinity of the oxygen atom is 33 kcal.mole^{-1} (for the addition of one electron). Explain why compounds involving the O^- ion are not observed.

10.7 Calculate the heat of formation of the hypothetical ionic compound, CaO_2, containing Ca^{2+} and O^- ions, assuming an equilibrium interionic distance of 0·22 nm and a Madelung Constant of 2·51939. The essential data are given in problems (10.4) and (10.6). Compare your answer with the heat of formation of CaO (-152 kcal.mole^{-1}) and with that of the actual compound, CaO_2, containing Ca^{2+} and O_2^{2-} ions ($-157\cdot5$ kcal.mole^{-1}).

270 **Atomic and Molecular Structure**

10.8 Plot the melting points (given in °C) of the following compounds against their molecular weights:

NaCl	800	CaF_2	1392	MgO	2642
$BeCl_2$	405	$CaBr_2$	730	CaO	2707
$MgCl_2$	712	CaI_2	575		
$CaCl_2$	772	KCl	776		
$SrCl_2$	872	RbCl	715		
$BaCl_2$	960	CuCl	422		
		AgCl	455		
		AuCl	170		

Bearing in mind your conclusions from problem (10.5) and the ideas concerning the cohesive forces of solids (Section 10.2), comment on the variation of melting point with (a) cation size, (b) anion size, (c) cation charge, (d) anion charge, and (e) nature of cation—main group of transition element. Suggest a qualitative interpretation of your conclusions in terms of (i) the polarizing (deforming) powers of the cations and (ii) the polarizabilities of the anions. Polarization in this instance may be taken to be the extent to which the electrons involved in producing the ions of the system are *shared* between the two types of atom.

10.9 To a certain degree of approximation, ions may be regarded as hard spheres. They may be assigned crystal radii, r_+ and r_-. Assuming that cations and anions are in contact in any crystalline arrangement, calculate the value of the radius ratio, r_-/r_+, corresponding to anion–anion contact for (a) a cation surrounded octahedrally by six anions, and (b) a cation surrounded tetrahedrally by four anions. These are the limits of stability for 6:6 and 4:4 lattices respectively. What factors determine the coordination numbers of anions in such compounds?

11

Molecules and Ions Containing Transition Elements

11.1 Introduction

The molecules considered in previous chapters have been those involving the main group elements. These have ground state electronic configurations in which the highest energy electrons do not occupy any d orbitals. The highest energy electrons are those responsible for the chemistry of an element and the orbitals in which they exist are called the valency orbitals. The main group elements possess valency shell configurations which may be written as $ns^{1\,\text{or}\,2}\,np^{0\,\text{to}\,6}$. Any d electrons which may be present (as for example in tin, (Kr) $5s^2\,4d^{10}\,5p^2$) are considerably more stable (exist at lower energy levels) than the s and p electrons which have greater values of the principal quantum number.

This is not so with the transition elements which possess none, one, or two s electrons in addition to a varying number of d electrons which are all of approximately the same energy. The $4s$ and $3d$, $5s$ and $4d$, and $6s$ and $5d$ sets of orbitals are, for the appropriate transition elements, of similar energies. Evidence of this is shown by the rather irregular filling of the d levels in which spin correlation effects predominate (Section 6.3).

The transition elements exist in several *oxidation states*. The latter term may be understood and defined by assuming that the oxidation state of atomic elements is zero, and if in the course of an oxidation process (oxidation can be defined as the loss of electrons by a species) n electrons are lost, as in equation (11.1),

$$M \rightarrow M^{n+} + ne^- \qquad (11.1)$$

then the oxidation state of the element in the M^{n+} ionic state is defined as $+n$. If electrons are gained by an element (corresponding to a reduction process), as in equation (11.2),

$$M + me^- \rightarrow M^{m-} \qquad (11.2)$$

the oxidation state produced is $-m$.

The oxidation states of ions produced by dissolving some transition elements in dilute acid are shown in Table 11.1.

Table 11.1. Oxidation States of Transition Elements Dissolved in Dilute Acid

Element	Oxidation state in dilute aqueous acidic solution
Fe $(s^2 d^6)$	Fe^{II} (d^6)
Co $(s^2 d^7)$	Co^{II} (d^7)
Ni $(s^2 d^8)$	Ni^{II} (d^8)
Zn $(s^2 d^{10})$	Zn^{II} (d^{10})
Cr $(s^1 d^5)$	Cr^{III} (d^3)

In the case of chromium, three electrons are removed from the ground state configuration $(s^1 d^5)$ in giving the chromium(III) ion. Two of these electrons are from d orbitals.

The aerial oxidation of iron(II) ion to iron(III) ion in aqueous solution,

$$2\text{Fe}^{II} + O_2 + 4H^+ \rightarrow 2\text{Fe}^{III} + 2H_2O \qquad (11.3)$$

involves the change, $d^6 \rightarrow d^5$, in the electronic configuration of the iron atom.

The permanganate ion, MnO_4^-, may be regarded ionically as $Mn^{7+} + 4O^{2-}$, i.e. a central manganese(VII) ion surrounded by four oxide ions. If this is so then the central manganese ion will have the d electronic configuration d^0, since the metal atom itself has the configuration $s^2 d^5$.

Table 11.2. The Oxidation States of the First Row Transition Elements

Element	Observed oxidation states
Sc	+3
Ti	+3, +4
V	+2, +3, +4, +5
Cr	+1, +2, +3, +6
Mn	+1, +2, +3, +4, +5, +6, +7
Fe	+1, +2, +3, +4, +6
Co	+1, +2, +3, +4
Ni	+1, +2, +3, +4
Cu	+1, +2
Zn	+1, +2

Molecules and Ions Containing Transition Elements

A summary of the various oxidation states of the first row transition elements is shown in Table 11.2.

There are a vast number of compounds involving the transition elements which are called *complexes*. These can be formulated as ML_6 or ML_4 depending on the stoichiometry where M = metal and L = ligand. Other stoichiometries of complexes are known but are not dealt with here. The ligand may be a single ion such as chloride, Cl^-, or it may be a group of atoms, e.g. NH_3, H_2O or pyridine. The charge upon the complex depends upon the charge upon the ligands (if any) and on the charge or oxidation state of the central metal atom. The ligands, Cl^-, NH_3, H_2O and pyridine, are *monodentate*, i.e. they form only one linkage to the metal atom. Some ligands can form more than one such linkage. For example, ethylenediamine, $NH_2CH_2CH_2NH_2$ (en), is *bi*dentate and can form a *chelate* ring by using the lone pair electrons of both nitrogen atoms in bond formation to a central metal atom, as shown in Figure 11.1. The bonds may be formally represented as being coordinate ones.

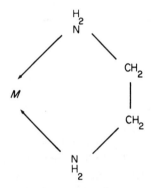

Figure 11.1. A chelate ring formed by the linkage of the nitrogen atoms of ethylenediamine to a metal atom

Due to the involvement of the d orbitals the Sidgewick–Powell ideas may not be applied to the stereochemistry of compounds of the transition elements and a more sophisticated approach is necessary.

There are two main theories which may be applied to transition element compounds and ions. These are the Crystal Field Theory and the Ligand Field Molecular Orbital Theory. They are used in a discussion of the following topics relating to transition element compounds and ions.

Stereochemistry

In the majority of the compounds and ions the coordination number (Section 10.5) of the transition element is observed to be either four or six. The four-coordination may be either *tetrahedral* or *square planar*, and the six-coordination is either regular *octahedral* or *tetragonal*.

The tetragonal structure involves four short square planar bonds and two longer axial bonds. Examples of compounds and ions possessing these four types of shape are shown in Table 11.3.

Table 11.3. The Shapes of Some Transition Element Compounds and Ions

Shape	Figure 9.1	Examples
Tetrahedral	(b)	$Ni(CO)_4$, $FeCl_4^-$, $Zn(NH_3)_4^{2+}$, $CoCl_4^{2-}$, MnO_4^-
Square planar	—	$Ni(CN)_4^{2-}$, $PtCl_4^{2-}$
Octahedral	(d)	$Fe(CN)_6^{4-}$, $Fe\ phen_3^{3+}$, $Co(NH_3)_6^{2+}$, $Fe\ Ox_3^{3-}$ [b]
Tetragonal [a]	—	$CuCl_2$, K_2CuF_4, CrF_2

[a] In this case the stereochemistry refers to the coordination of the central metal ion in the solid compounds.
[b] phen = *o*-phenanthroline, and Ox = oxalate.

Absorption Spectra

An absorption spectrum is obtained by determining the absorbance (Section 2.4) of a solution of a compound in a suitable solvent at various wavelengths. The absorbance versus wavelength (or frequency) plot is the absorption spectrum. In general, with covalent compounds the electrons may be considered to exist in either sigma or pi bonding orbitals or in non-bonding sigma orbitals with π^* orbitals being infrequently occupied (e.g. O_2, NO). The σ^* and the π^* orbitals are normally vacant. A general order of energies of these orbitals is indicated in Figure 11.2.

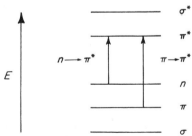

Figure 11.2. A general order of energies for the σ, π, n, π^* and σ^* levels

Molecules and Ions Containing Transition Elements

Consider the general case where the σ, π and n levels are occupied and the π^* and σ^* levels are vacant. The lowest energy transition which corresponds to absorption of a quantum of suitable frequency may be labelled $n \to \pi^*$. This is an *electronic transition* from the highest energy occupied orbital (n) to the orbital of next highest energy (π^*), which is normally vacant. Such transitions occur in compounds containing lone pair electrons, e.g. formaldehyde and acetone, at frequencies corresponding to the ultraviolet region of the spectrum. The higher energy transitions which are possible ($n \to \sigma^*$, $\pi \to \sigma^*$, $\sigma \to \sigma^*$) occur at higher frequencies and are usually apparent in the far ultraviolet region (below 200 nm). It is, of course, possible for electronic transitions to occur in the visible region, thus producing absorption in that region and consequently being responsible for the *colour* of the compound.

In the majority of covalent molecules of the non-transition elements, however, the lowest energy transitions are in the ultraviolet region of the spectrum and the compounds are in consequence not coloured.

The situation with compounds and ions containing transition elements is very different. Coloured compounds are the rule rather than the exception. That the compounds are coloured is of no relevance in itself. The main fact is that the compounds in general absorb at longer wavelengths than those of non-transition elements.

Sizes of Ions: Hydration and Lattice Energies

The majority of the first row transition elements form doubly charged positive ions. These ions may exist in crystalline materials and they may have a crystal radius assigned to them. A graph showing the variation in the radii of the $+2$ ions with increasing atomic number is shown in Figure 11.3. Since the hydration energy (the energy released in the reaction $M^{2+}_{(g)} \to M^{2+}_{(aq)}$) and the lattice energy (Section 10.5) (of, say, the fluorides) depend upon the radius of the ion concerned, we might expect these quantities to vary similarly to the radii with atomic number. This is so, as may be seen from Figures 11.4 and 11.5.

In a series of ions of identical charge it would be expected that as the nuclear charge (Z) increased the ionic size would decrease. This occurs in the series of $+3$ ions of the lanthanide elements as shown in Figure 11.6.

Magnetic Properties

In general, there are two types of magnetic properties which are important in discussing the electronic configurations of the transition element compounds and ions. These are *diamagnetism* and *paramagnetism*.

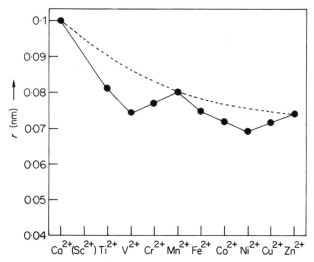

Figure 11.3. The variation in the radius of doubly charged ions of the first transition series

Diamagnetic substances contain no unpaired electrons and their interaction with a magnetic field is to move to a region of minimum field strength. A paramagnetic substance possesses one or more unpaired

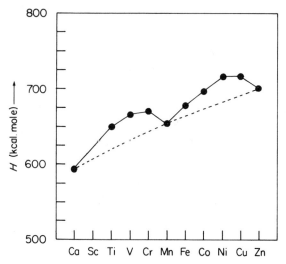

Figure 11.4. The variation of the hydration energies of the doubly charged cations of the first transition series

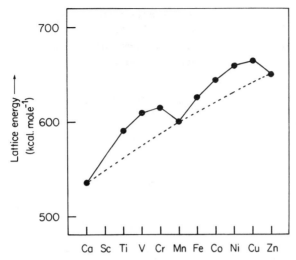

Figure 11.5. The variation of the lattice energies of the divalent fluorides of the first transition series

electrons and tends to concentrate the lines of force and takes up a position in a magnetic field where the field strength is a maximum.

If an amount of a substance is weighed using a magnetic balance (Gouy balance) it is heavier in the applied magnetic field than in the absence of the field if it is *paramagnetic*. Diamagnetic substances are pushed out of the field so that they weigh less in the presence of a field than in its absence.

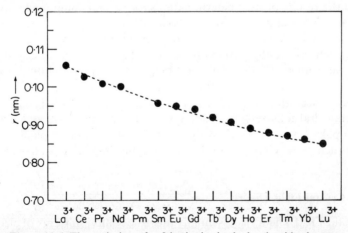

Figure 11.6. The variation of +3 ionic size in the lanthanide elements

In the accurate determination of the paramagnetism of a substance, it is necessary to allow a correction for the diamagnetic contribution of the paired electrons in the system. From these measurements the number of unpaired electrons in a molecule may be determined. In Table 11.4 are

Table 11.4. The Numbers of Unpaired Electrons in Some Octahedral Complexes as Indicated by Magnetic Measurements

Compound	Electronic configuration of metal ion	Number of unpaired electrons
$Cr(dipy^a)_3Br_2$	d^4	2
$Mn(acac^b)_3$	d^4	4
$K_4Mn(CN)_6$	d^5	1
$K_3Fe(CN)_6$	d^5	1
$Mn(py^c)_6Cl_2$	d^5	5
K_3FeF_6	d^5	5
$K_4Fe(CN)_6$	d^6	0
$Co(NH_3)_6Cl_3$	d^6	0
$Fe(NH_3)_6Cl_2$	d^6	4
K_3CoF_6	d^6	4
$K_4Co(NO_2)_6$	d^7	1
$Co(NH_3)_6Cl_2$	d^7	3

[a] dipy = α, α' dipyridyl.
[b] acac = acetylacetonate.
[c] py = pyridine.

shown the results of such determinations for several compounds involving transition metal ions. From these data it is clear that, depending upon the environment of the transition metal cation, there are two different electronic configurations possible. It has also been observed that metal ions with d^1, d^2 or d^3 configurations *always* form compounds involving one, two or three unpaired electrons respectively.

Other observations of importance are that compounds of nickel involving octahedral ions such as $Ni(NH_3)_6^{2+}$ are paramagnetic and possess two unpaired electrons, while the more common square planar complexes such as $Ni(CN)_4^{2-}$ are diamagnetic (no unpaired electrons).

11.2 Crystal Field Theory

Crystal Field Theory is essentially a consideration in terms of electrostatics of the effects of the various symmetrical methods of arranging ligands, considered as being negative point charges, around a central atom or ion.

The crystal field effects of octahedral, square planar and tetrahedral coordination upon the central atom or ion will be dealt with. As far as transition elements are concerned a consideration of the crystal field effects upon the d orbitals is of supreme importance.

The d orbitals of an atom are normally five-fold degenerate (Section 5.1) and differ in their spatial orientations (Figure 5.7). They may be divided into two groups, the one being the d_{z^2} and $d_{x^2-y^2}$ orbitals which are primarily directed along the coordinate axes (x, y and z), and the other being the d_{xy}, d_{xz} and d_{yz} orbitals which are directed along axes which bisect the coordinate axes.

These two groups of d orbitals may be labelled in the following way:

$d_{z^2}, d_{x^2-y^2}$	d_γ	e_g
d_{xy}, d_{xz}, d_{yz}	d_ε	t_{2g}

the symbol, e_g, being the group theory symbol for a doubly degenerate group of orbitals which are symmetric to inversion, and t_{2g} being that for a triply degenerate group (also symmetric to inversion). The d_γ and d_ε labels are sometimes used for convenience.

If a system which, say, possesses one electron in each of the five d orbitals is surrounded by an octahedral arrangement of six negative charges (an octahedral crystal field), such that the negative charges approach the central atom or ion along the coordinate axes, the orbital degeneracy is split. The electrons in the d orbitals will experience repulsion from the six negative point charges and the d electron energies will be increased. The electrons in the d_{z^2} and $d_{x^2-y^2}$ orbitals, however, will experience greater repulsion, since they are directed along the coordinate axes, than the electrons in the other group of orbitals. The difference in energy between the e_g and t_{2g} levels is given the symbol, Δ_0, and is known as the octahedral crystal field splitting energy.

This situation is summarized in Figure 11.7. It must be emphasized that the overall effect of the crystal field would be to confer stability upon the system. Figure 11.7 deals only with the effect of the crystal field upon the d electron energies and not with the stabilization of the whole system as a result of the interaction of the crystal field with the nuclear charge of the central metal atom.

If the six localized ligand charges were spread out over the surface of a

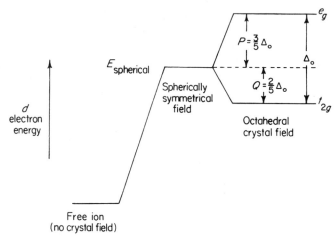

Figure 11.7. The effects upon d orbital energies of a spherical and an octahedral crystal field

sphere with the metal ion at its centre, the d orbital energy would be given by $E_{\text{spherical}}$. Since the charges, whatever their spatial distribution, must produce in total the same repulsive effect upon the energy of the symmetrically filled d orbitals, we can write (for the five electrons, one in each orbital):

$$5E_S = 2(E_S + P) + 3(E_S - Q) \qquad (11.4)$$

where E_S is the energy of a d electron in a spherically symmetrical crystal field, P is the destabilization energy produced by the octahedral crystal field upon the e_g orbitals and Q is the stabilization energy of the t_{2g} orbitals in the same field. The distribution of the field does not alter the total d energy of a d^5 configuration in which each orbital contains one electron. (The same is true for two electrons per d orbital.)

We may also express Δ_0 in terms of P and Q:

$$\Delta_0 = P + Q \qquad (11.5)$$

and equation (11.4), after multiplying out the factors on the right-hand side, becomes:

$$5E_S = 2E_S + 2P + 3E_S - 3Q \qquad (11.6)$$

and

$$2P + 3Q = 0 \qquad (11.7)$$

Substitution of the value of P given by equation (11.5), $\Delta_0 - Q$, into equation (11.6) gives

$$2(\Delta_0 - Q) - 3Q = 0 \qquad (11.8)$$

or

$$2\Delta_0 - 5Q = 0 \qquad (11.9)$$

so that

$$Q = \tfrac{2}{5}\Delta_0 \qquad (11.10)$$

and it follows that

$$P = \Delta_0 - \tfrac{2}{5}\Delta_0 = \tfrac{3}{5}\Delta_0 \qquad (11.11)$$

These values of P and Q are indicated in Figure 11.7.

The stereochemistry, absorption spectra, ion sizes and magnetic properties of the species under discussion are intimately related to the d electronic configurations of the metal ion or atom. The d electronic configuration of an atom or ion with a given number of d electrons depends on the relative magnitudes of two energies. These are Δ_0 (the difference in energy between the e_g and t_{2g} levels) and the pairing energy, E_P, which has to be overcome when two electrons occupy the same orbital. E_P is composed of the charge correlation occurring when two electrons come together (with opposed spins) in one orbital.

For metals with one, two, or three d electrons the value of Δ_0 does not affect the electronic arrangement since only the lower energy, t_{2g}, level will be involved. If the metal has more than three d electrons then the relative sizes of Δ and E_P determine the ground state electronic configuration.

If E_P is *greater* than Δ_0 we may assume that the crystal field is *weak*, while if E_P is *smaller* than Δ_0 we may assume it to be *strong*. In this way we can divide crystal field effects into weak and strong categories. The strength of the crystal field affects the d configurations of those metal atoms or ions possessing four, five, six or seven d electrons.

Consider the d^4 configuration. If E_P is larger than Δ_0 then the fourth electron will occupy one of the e_g orbitals in preference to pairing with one of the electrons already present in the t_{2g} orbitals. If, on the other hand, Δ_0 is larger than E_P there will be *spin pairing* and the fourth electron will pair up to give the configuration t_{2g}^4 rather than $t_{2g}^3 e_g^1$ as in the *spin-free* case.

Using similar arguments it is possible to derive the electronic configurations of various numbers of d electrons, these being shown in Table 11.5.

Table 11.5. Ground State d Electronic Configuration in Weak and Strong Octahedral Crystal Fields

Number of d electrons	Weak crystal field ($E_P > \Delta_0$)		Strong crystal field ($\Delta_0 > E_P$)	
	t_{2g}	e_g	t_{2g}	e_g
1	1	0	1	0
2	2	0	2	0
3	3	0	3	0
4	3	1	4	0
5	3	2	5	0
6	4	2	6	0
7	5	2	6	1
8	6	2	6	2
9	6	3	6	3
10	6	4	6	4

The effects of octahedral crystal field strength upon the number of unpaired electrons are summarized in Table 11.6.

Table 11.6. Effect of Octahedral Crystal Field Strength upon the Numbers of Unpaired Electrons in a given d Configuration

Number of d electrons	Number of unpaired d electrons	
	Weak field	Strong field
1	1	1
2	2	2
3	3	3
4	4	2
5	5	1
6	4	0
7	3	1
8	2	2
9	1	1
10	0	0

The predictions of crystal field theory as shown in Table 11.6 for the numbers of unpaired electrons in a given d configuration are made from the data in Table 11.5, bearing in mind the operation of Hund's Rules (Section 6.3) in considering the filling of the t_{2g} and e_g levels. They may be compared with the results of magnetic measurements which are shown in Table 11.4.

Distorted Octahedral Ions

In addition to ions which are regularly octahedral, there are also those which are distorted from octahedral symmetry, examples of these ions being shown in Table 11.3. One of the distortions observed is the tetragonal distortion in which a complex ion contains four short and two long bonds or metal–ligand distances.

Such tetragonal distortion of an octahedral complex may be dealt with in terms of a reconsideration of the crystal field theory for a regular octahedral complex. This latter theory has been worked out so far in terms of the crystal field effect of a regular octahedral distribution of six ligands (or point charges) upon the d orbital energies in a d^5 configuration, i.e. in the case where each orbital contains a single electron. The same treatment applies to a d^{10} configuration in which each orbital is full. The conclusion is again that the five-fold degeneracy of the d orbitals is split into the two-fold and three-fold degenerate e_g and t_{2g} levels respectively.

The simple treatment is only accurate for d^5 (spin-free) and d^{10} configurations. In all other cases where there are less symmetrical fillings of the d orbitals other effects must be taken into consideration.

Asymmetric filling of the t_{2g} level may influence the shape of an otherwise regular octahedral complex, but asymmetry in the e_g level has much more effect. The reason for this lies in the fact that the e_g orbitals are spatially directed towards the ligands while the t_{2g} orbitals bisect the ligand directions.

Let us consider the stereochemical aspects of the asymmetrical filling of the e_g level. Such d electronic configurations which are asymmetric as far as the e_g level is concerned are $t_{2g}^3 e_g^1$, $t_{2g}^6 e_g^1$ and $t_{2g}^6 e_g^3$, in which there are an odd number of electrons in the e_g orbitals. The question is then, which is the more stable configuration, $d_{z^2}^1$ or $d_{x^2-y^2}^1$ in the case of the e_g^2 filling, and $d_{z^2}^2 d_{x^2-y^2}^1$ or $d_{z^2}^1 d_{x^2-y^2}^2$ in the case of the e_g^3 filling?

It is difficult to decide which of the two alternative configurations is the more stable in each case from theoretical considerations. Experiment provides the answer. If, in these asymmetric e_g configurations, the d_{z^2} orbital happened to be more stable than the $d_{x^2-y^2}$ orbital then the favoured states would be $d_{z^2}^1$ and $d_{z^2}^2, d_{x^2-y^2}^1$ in the two cases.

In each case there would be a greater electron density along the z coordinate axis than along the x and y axes. The attractive force operating between the central metal ion and the six ligands is offset by the presence of electrons in the d orbitals of the metal ion. In the case of the asymmetric filling of the e_g orbitals the reduction in the attractive force would be greater in the z direction than along the other axes. The consequence of this would be a complex containing four short bonds (in the xy plane) and two long bonds (along the z axis).

If the asymmetry in the filling of the e_g orbitals were opposite to the case discussed above (i.e. the $d_{x^2-y^2}$ orbital being preferred to the d_{z^2} orbital) the opposite distortion would occur with the complex having four long bonds in the xy plane and two short bonds along the z axis.

These two distortions together with the corresponding e_g orbital energies are shown in Figure 11.8. The consequence of the distortions

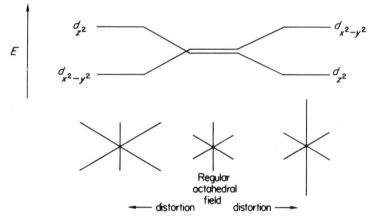

Figure 11.8. Effects of distortion of an octahedral crystal field upon the energies of the e_g orbitals

shown in Figure 11.8 is the breakdown of the degeneracy of the e_g level. The experimentally observed deviations from regular octahedral coordination in an ML_6 complex are generally of the four short bond, two long bond type, and it would appear that whenever the e_g level is asymmetrically filled that its degeneracy is split according to the right-hand side of the diagram in Figure 11.8. That is, the d_{z^2} orbital becomes stabilized and the $d_{x^2-y^2}$ orbital becomes destabilized with respect to the energy of the doubly degenerate e_g level in a regular octahedral field.

Such arguments correlate theory and observation (see Table 11.3) satisfactorily.

The square planar complexes which are common in the compounds containing nickel(II), palladium(II) and platinum(II) may be treated as an extreme form of tetragonal distortion in which two of the bonds (those along the z axis) are of infinite length. In the case of Ni^{II}, Pd^{II} and Pt^{II} ions there are two electrons to be disposed of in the e_g orbitals (there are six in the t_{2g} orbitals).

The two electrons in the e_g level may be distributed evenly ($d_{z^2}{}^1 d_{x^2-y^2}{}^1$), with their spins parallel producing a triplet state ($s_1 = s_2 = \frac{1}{2}$, $S = 1$, $2S+1 = 3$), the alternative states being the $d_{z^2}{}^1 d_{x^2-y^2}{}^1$ singlet ($s_1 = \frac{1}{2}$, $s_2 = -\frac{1}{2}$, $S = 0$, $2S+1 = 1$) and the $d_{z^2}{}^2$ and $d_{x^2-y^2}{}^2$ singlet states. In an octahedral field the triplet state, $d_{z^2}{}^1 d_{x^2-y^2}{}^1$, is the most stable (Hund's Rules) where the e_g levels are strictly degenerate. This is the case only in a regular octahedral field. The corresponding singlet level is at a higher energy. The energies of these states do not vary with the distortion from regular octahedral symmetry of the crystal field. As has been pointed out, however, the relative energies of the d_{z^2} and $d_{x^2-y^2}$ orbitals do depend on the extent of the tetragonal distortion and, if this proceeds in the four short, two long bond manner, then the $d_{z^2}{}^2$ configuration will become increasingly stable, the opposite being the case for the $d_{x^2-y^2}{}^2$ configuration. As is shown in Figure 11.9, at a certain distortion the $d_{z^2}{}^2$ and $d_{z^2}{}^1 d_{x^2-y^2}{}^1$ configurations become degenerate, and greater distortion still makes the $d_{z^2}{}^2$ state the ground state. Provided that the distortion is large (e.g. four short bonds and two very long bonds) the d_{z^2} configuration

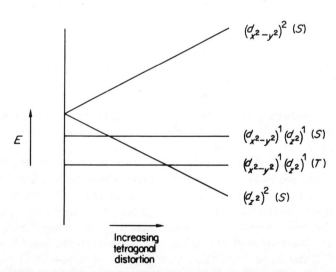

Figure 11.9. Effects of tetragonal distortion on the ground state of a d^8 ion

becomes the ground state. Thus, in the square planar Ni^{II}, Pd^{II} and Pt^{II} complexes the ground state is diamagnetic.

An e_g^2 configuration will be a triplet state in a regularly octahedral field (Hund's Rules). A small distortion breaks down the degeneracy of the e_g orbitals and, providing that the energy difference between them is smaller than the pairing energy, the triplet state will be the ground state. With larger distortions it is possible for the $d_{x^2-y^2} - d_{z^2}$ energy gap to be larger than the pairing energy and for spin-pairing in the d_{z^2} orbital to occur.

The breakdown of the degeneracy of the e_g level caused by distortion is an example of a theorem proposed by Jahn and Teller in 1937. This may be stated as follows:

> Any non-linear molecular system in a degenerate electronic state will be unstable and will undergo a distortion which will lower its symmetry and split the degenerate state.

The theorem does not indicate the nature of the distortion or its extent. The distortion of the regular octahedral systems referred to above is *not* caused by the breakdown of the degeneracy. The breakdown of the degeneracy is caused by the distortion which in turn is justified in terms of the energetics of the systems. Compared to the undistorted state (regular octahedral) an odd electron gains stability when the symmetry is lowered (i.e. distortion occurs), since its orbital energy is also lowered.

Absorption Spectra

The absorption by a molecular system of a quantum of visible light causes an electronic transition. The energy corresponding to the difference between the two levels concerned may be calculated. Consider a transition occurring at 480 nm. The quantum energy is calculated to be 60 kcal.mole^{-1} (251 kJ.mole^{-1}) by the method shown in Section 2.2.

The quantity Δ_0, if it corresponds to visible energies, may give very good grounds for the interpretation of absorption spectra of transition element complexes as consisting of d–d transitions. This means that only d electrons are involved and that a d electron in a lower level is transferred to an upper level upon absorption of sufficient quantum energy. On a simplified basis these transitions would be $t_{2g} \rightarrow e_g$. Since the value of l (2 for a d electron) does not change, a d–d transition is *forbidden* by the Laporte Rule (Section 6.1). This means that any of these transitions which occur at all do so with very low intensity. The molar absorption coefficient (Section 2.4, equation 2.15) is one measure of the *allowedness* of a transition. A fully allowed transition has an associated absorption coefficient in the region of 10^4 l.mole^{-1}.cm^{-1}. Several factors may operate to increase the forbiddenness of a transition. The absorption coefficients associated with d–d transitions are usually of the order of only 5 l.mole^{-1}.cm^{-1}.

The simplest d–d transition which may be observed is that involved in the spectrum of the Ti^{III} (H$_2$O)$_6^{3+}$ ion. The spectrum of this species is shown in Figure 11.10. In the visible region there is a broad asymmetric absorption band which corresponds to the d–d transition.

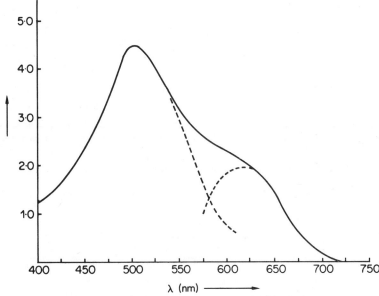

Figure 11.10. The spectrum of the Ti^{III} (H$_2$O)$_6^{3+}$ ion

The d–d transition in the case of Ti^{III} may be designated as a $t_{2g} \to e_g$ transition. There is only one d electron involved and the excited state configuration is e_g^1. Under these conditions we would expect the Jahn–Teller effect to operate and split the degeneracy of the e_g orbitals such that there may be then two transitions instead of one. These two transitions may be written as:

$$t_{2g} \to d_{z^2}$$

and

$$t_{2g} \to d_{x^2-y^2}$$

and would be expected to occur at different wavelengths. This is the probable explanation of the asymmetry of the d–d absorption band which is a composite band in which the two d–d transitions overlap each other. This is indicated by the dotted lines in Figure 11.10.

The blue colour of the $Cu(H_2O)_6^{2+}$ ion is due to the transition, $t_{2g}^6 e_g^3 \rightarrow t_{2g}^5 e_g^4$, which occurs in the red region of the visible spectrum. Again the transition is affected by the Jahn–Teller effect which in this case operates mainly on the ground state e_g^3 configuration.

The absorption maximum is the spectrum of $Ti^{III}(H_2O)_6^{3+}$ occurs at 500 nm. This corresponds to a quantum energy of 57·4 kcal.mole^{-1} (240 kJ.mole^{-1}). In such a way we may calculate the value of Δ_0 which represents the difference in energy between the e_g and t_{2g} levels.

Although the spectra of complex ions in which there are more than one d electron are more complicated than that of the d^1 case just discussed, it is possible to derive a value for Δ_0 in each case.

For example, we may consider a series of different ligands and their effects upon Δ_0. In the case of the chromium(III) ion which has a d^3 configuration, the values of Δ_0 (estimated from measurements of absorption spectra) in the presence of different ligands are shown in Table 11.7.

Table 11.7. The Variation of Δ_0 with the Nature of the Ligands in Cr^{III} Complexes

Ligands	Δ_0	
	(kcal.mole^{-1})	(kJ.mole^{-1})
6Cl$^-$	39·0	163
6H$_2$O	50·0	210
6NH$_3$	62·0	260
6CN$^-$	75·5	316

The figures in Table 11.7 may be interpreted in terms of the crystal field strengths exerted by the sets of ligands upon the central metal ion, and hence the ligands may be arranged in order of increasing crystal field strength, viz.:

$$Cl^- < H_2O < NH_3 < CN^-$$

Similar procedures have been carried out with a great variety of ligands and with a whole series of different metal ions, and, in general, it is possible to write down what is known as the *spectrochemical series*, viz.:

$$I^- < Br^- < Cl^- < F^- < H_2O < NH_3$$
$$< en < dipy < phen < NO_2^- < CN^- = C_2H_4 = CO$$

Molecules and Ions Containing Transition Elements

The spectrochemical series is ordered such that the ligands appear in order of increasing crystal field strength as measured by the magnitude of Δ_0. Of course, the value of Δ_0 does depend upon the particular metal ion involved in the complex as the data in Table 11.8 indicate.

Table 11.8. The Variation of Δ_0 with the Metal Ion involved in $M(H_2O)_6^{3+}$ Complex Ions

Metal ion	d Configuration	Δ_0 (kcal.mole^{-1})	Δ_0 (kJ.mole^{-1})
Ti^{III}	d^1	57·4	241
V^{III}	d^2	50·8	212
Cr^{III}	d^3	49·9	209
Mn^{III}	d^4 (spin-free)	60·3	253
Fe^{III}	d^5 (spin-free)	39·3	164

The spectrochemical series is referred to again in Section 11.4 where the order of the ligands is discussed at greater length. As far as *crystal field theory* is concerned, it is sufficient to point out that it explains why there should be d–d spectral transitions. Since in crystal field theory we are considering ligands to be point charges, it is not possible to deal in detail with the differences between ligands as are indicated by the spectrochemical series. These latter have to be dealt with by using the more complicated ligand field molecular orbital approach.

The value of the crystal field splitting parameter, Δ_0, does vary with the oxidation state of the central metal atom of a complex. Data exemplifying this point are shown in Table 11.9 for some hexaquo complexes.

Table 11.9. The Variation of Δ_0 with the Oxidation State of the Metal Atom

Metal	Δ_0 for $M(H_2O)_6$			
	+2 state		+3 state	
	(kcal.mole^{-1})	(kJ.mole^{-1})	(kcal.mole^{-1})	(kJ.-mole^{-1})
Cr	40·5	170	49·9	209
Fe	27·1	114	39·3	164
Co	23·5	98	52·3	220

Generally it may be concluded that the increasing oxidation state of a metal, resulting in an increased effective nuclear charge, produces complexes with shorter bonds so that crystal field effects are increased.

Size Effects

It now remains to discuss the influence of crystal field theory upon the relative sizes of transition metal cations and the quantities (hydration energies and lattice energies) which are affected by them.

Consider the sizes of the +2 ions of the first row transition elements (Ti → Zn). It is of interest to compare their sizes with that of the calcium cation, Ca^{II}, which does not contain any d electrons and is, in consequence, not affected by the crystal field of the surrounding ligands in the crystalline state. In comparing the size of the Ti^{II}, d^2, ion with that of Ca^{II} we have to note that the increasing nuclear charge (from 20 to 22) would be expected to produce a contraction, such contraction only being offset by the shielding of the ligands from the nuclear charge by the additional electrons. Crystal field theory indicates that the two d electrons in Ti^{II} would occupy the t_{2g} level and thus would not contribute very much to the shielding of the ligands from the nuclear charge. (The t_{2g} orbitals are directed so as to bisect the metal–ligand directions.) The crystal field effect upon the radius of the Ti^{II} cation would be to allow the nuclear charge to be more effective than in the absence of such a field and, in consequence, the ion is smaller than would be normally expected. In going to the V^{II} ion which has a d^3 configuration the extra electron again would occupy one of the t_{2g} orbitals and this would lead to a smaller ion than expected since the extra electron would not contribute to very efficient shielding of the nuclear charge.

In the case of the Cr^{II} cation (d^4) the extra electron (compared to V^{II}) occupies one of the e_g orbitals. In this case the extra electron is in a position (along the coordinate axes or the metal–ligand directions) to contribute to very efficient shielding of the nuclear charge. Here, as can be seen from Figure 11.3 the effect is large enough to offset the nuclear charge to an extent which produces a larger ion than V^{II}, in spite of the increased nuclear charge.

With the high spin, Mn^{II}, d^5, cation there are two electrons in the more efficiently shielding e_g orbitals and again there is an increase in the crystal radius of the Mn^{II} cation. In this special case, with all the five d orbitals singly occupied, crystal field effects are cancelled out since the three electrons in the t_{2g} level are $3 \times \frac{2}{5}\Delta_0 = \frac{6}{5}\Delta_0$ more stable than in the case of a spherically symmetrical crystal field and the two electrons in the e_g orbitals are $2 \times \frac{3}{5}\Delta_0 = \frac{6}{5}\Delta_0$ less stable. The overall *crystal field stabilization energy* is zero.

The radii of the ions Mn^{II} to Zn^{II} follow the same pattern as do those from Ca^{II} to Cr^{II} for similar reasons, and in Zn^{II} crystal field effects are again zero.

Molecules and Ions Containing Transition Elements

Crystal field theory gives a very satisfactory explanation of the variation of the crystal radii of the spin free, doubly charged cations of the first row transition elements. Similar reasonings can be applied to the other rows and to other oxidation states with similar success.

Since the lattice energy of a crystal is inversely proportional to the interionic distance (Section 10.5), an unusually small metal ion causes an unusually high lattice energy. Thus the variations in lattice energy as shown in Figure 11.5 may be explained.

The hydration energy of an ion is inversely proportional to its crystal radius and hence the data in Figure 11.4 are understandable.

For completeness and accuracy it should be pointed out that in the case of d^4 (spin-free) and d^9 configurations the Jahn–Teller effect operates and distortions from octahedral symmetry are to be expected. These in turn affect the concept of a crystal radius since such ions are not surrounded by ligands at the same distances.

Summary

From the above discussion it appears that a simple consideration of crystal field theory can explain qualitatively:

(1) the electronic configurations of transition metal ions containing d electrons and, in consequence,
(2) the magnetic properties of such ions,
(3) the d–d absorption spectra of the ions, and
(4) the variations in crystal radii and hydration energies of their ions and the lattice energies of their compounds.

11.3 Crystal Field Theory of Tetrahedral Complexes

The surrounding of a central metal ion or atom with a tetrahedral crystal field may be thought to occur by placing the four negative point charges (ligands) at the alternate corners of a cube whose face centres coincide with points on the coordinate axes, x, y and z. This situation is shown in Figure 11.11.

The d orbitals have their usual spatial distribution in Figure 11.11, the e set being directed along the coordinate axes (towards the face centres of the cube) and the t_2 set bisecting the coordinate axes (the g subscripts are dropped when considering tetrahedral fields, since the tetrahedral point group does not contain an inversion centre). The four tetrahedrally disposed ligands affect the d electron energies unequally since the metal–ligand directions form an angle of 54° 44′ with the coordinate axes along which the e orbitals are directed and only 35° 16′ with the directions of the

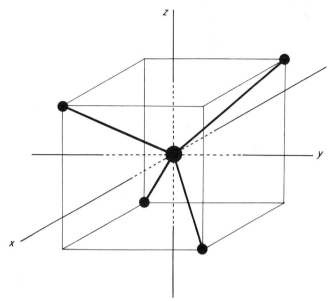

Figure 11.11. Showing the relationship between a tetrahedrally coordinated central point and the coordinate axes

t_2 orbitals. In consequence, the t_2 energies are destabilized with respect to the spherically symmetrical field and the e energies are stabilized. The five-fold degeneracy of the d orbitals is split but in the opposite sense to the case of an octahedral field. The breakdown in degeneracy is shown in Figure 11.12.

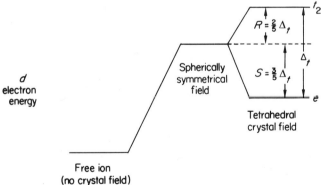

Figure 11.12. The effects of a spherical and a tetrahedral crystal field upon d orbital energies. (From page 871 of *Advanced Inorganic Chemistry*, by F. A. Cotton and G. Wilkinson, Interscience, 1966; reprinted with the permission of the publishers)

Molecules and Ions Containing Transition Elements

Compared to the energy of the d electrons in a spherically symmetrical field the e orbitals are stabilized by an amount, S, and the t_2 orbitals are destabilized by an amount, R, when a tetrahedral field is applied. Considering the case of a d^{10} configuration where crystal field effects cancel out, we can write:

$$4S = 6R \tag{11.12}$$

and if

$$R + S = \Delta_t \tag{11.13}$$

then

$$R = \Delta_t - \tfrac{3}{2}R$$

or

$$\tfrac{5}{2}R = \Delta_t$$

or

$$R = \tfrac{2}{5}\Delta_t \tag{11.14}$$

so that

$$S = \tfrac{3}{5}\Delta_t \tag{11.15}$$

Whether any particular d electron configuration is spin-free or spin-paired depends upon the relative magnitudes of Δ_t and the pairing energy. It is of great significance to compare the values of Δ_t with Δ_0 for an octahedral crystal field. Δ_t is less than Δ_0 for two reasons, the first being that only four ligands are involved in the production of Δ_t compared to six in the octahedral case. The second reason is that the ligands have different spatial orientations with respect to the d orbital directions. The electrostatic repulsive effects of negative point charges directed along the metal–ligand bonds and at the same distance from the metal atom as in the octahedral case may be calculated, the result being that

$$\Delta_t = \tfrac{4}{9}\Delta_0 \tag{11.16}$$

The consequences of this relationship are dependent upon a comparison of Δ_t with the pairing energy of two electrons. The pairing energy is always larger than Δ_t and tetrahedral spin-paired complexes never occur. There is no fundamental theoretical reason for this being so. To get spin-pairing Δ_t would have to be greater than the pairing energy and this may only be achieved by altering the stereochemistry (and the coordination number) of the complex. For the possible spin-free tetrahedral complexes the crystal field stabilization energies (CFSE) associated with various d electron configurations are shown in Table 11.10, together with the values obtained in an octahedral crystal field.

Table 11.10. CFSE Values for Spin-free Tetrahedral and Octahedral Configurations

Number of d electrons	CFSE	
	Tetrahedral	Octahedral
0, 5, 10	0	0
1, 6	$\frac{3}{5}\Delta_t = 0.267\Delta_0$	$0.4\Delta_0$
2, 7	$\frac{6}{5}\Delta_t = 0.534\Delta_0$	$0.8\Delta_0$
3, 8	$\frac{4}{5}\Delta_t = 0.355\Delta_0$	$1.2\Delta_0$
4, 9	$\frac{2}{5}\Delta_t = 0.178\Delta_0$	$0.6\Delta_0$

The CFSE values for tetrahedral coordination are less than those for octahedral coordination except for the d^0, d^5 and d^{10} cases when the CFSE is zero in both cases. In consequence, tetrahedral coordination is not common. It occurs in the d^0, d^5 and d^{10} cases and in very few others where the difference in CFSE between octahedral and tetrahedral symmetry is not great. A negative ligand exerting a low crystal field with no possibility of double bonding sometimes can be the factor deciding the stereochemistry in any particular case. Examples of tetrahedral species are $TiCl_4$, $ZrCl_4$, MnO_4^- (all d^0), $FeCl_4^-$ (d^5) and $ZpCl_4^{2-}$ (d^{10}). Others are $VCl_4(d^1)$, VCl_4^-, FeO_4^{2-} (d^2) and $CoCl_4^{2-}$ (d^7).

In the formation of a complex there is a tendency for the positive charge on the metal ion to be neutralized by the donation of electronic charge from the ligand atoms or groups (the Pauling Electroneutrality Principle). This tendency largely determines the stoichiometry of the complex. Ligands which are good donors (low ionization potentials) will tend to form complexes of lower ligand/metal ratios than those which are not as good.

The factors which influence the stereochemistry of an ML_4 complex are (1) the crystal field stabilization energy, and (2) the residual charge on the ligand. The polarizable ligands, Cl^- and O^{2-}, can act as good donors and produce ML_4 complexes, and the ligand–ligand repulsion can cause the complex to be tetrahedral provided that factor (1) is favourable. This latter condition is fulfilled only in the few cases mentioned above.

A further consequence of tetrahedral coordination is that complexes with this stereochemistry have absorption spectra which are red-shifted

Molecules and Ions Containing Transition Elements

(i.e. to longer wavelengths) with respect to those of octahedrally coordinated metals. This is due to the relation (11.16), and is a direct consequence of the difference in the Δ values in the two cases. It is best exemplified by comparing the spectra of Co^{II} in the two environments. Figure 11.13 shows the spectra of the octahedral $Co(H_2O)_6^{2+}$ and $CoCl_4^{2-}$ complexes.

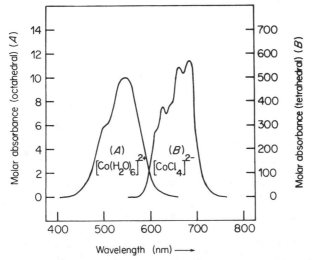

Figure 11.13. The d–d spectra of $Co(H_2O)_6^{2+}$ and $CoCl_4^{2-}$ (After Cotton and Wilkinson, *Advanced Inorganic Chemistry*, Interscience, 1966)

11.4 Ligand Field Molecular Orbital Theory

The crystal field theory of complexes as outlined in the preceding sections of this chapter attaches no significance to any metal–ligand interaction other than that expected for point charges. No indication of the type of bonding is given. If, in fact, there was no mixing of the ligand and metal orbitals (and therefore, electrons) this would be detectable by experiment.

Compounds with unpaired electrons (which may have $s = \pm\frac{1}{2}$) show *electron spin resonance* (e.s.r.) spectra in which absorption of radio frequency quanta cause a reversal of the spin state of an electron. In a magnetic field the spins of the unpaired electrons may be parallel to the applied field or antiparallel to it. Absorption of radiation can cause a reversal of a spin and give rise to the e.s.r. spectrum of the compound. Such a transition is illustrated in Figure 11.14, in which the spin parallel to the field is 'flipped' by absorption of the requisite amount of energy

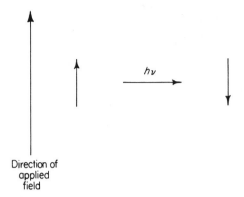

Figure 11.14. The absorption of a quantum causing an electron spin reversal

to become antiparallel to the field. The energy gap between the two states is dependent upon the magnitude of the applied field. Figure 11.15 illustrates this. As is shown in Figure 11.15, the energy of the transition is proportional to the applied field strength, H:

$$hv \propto H \qquad (11.17)$$

The experimental method of determining e.s.r. spectra is to irradiate the sample with a fixed frequency radiation (10 MHz) and to alter the strength of the applied magnetic field until the absorption condition is fulfilled. A spectrum may consist of a plot of absorption against the field

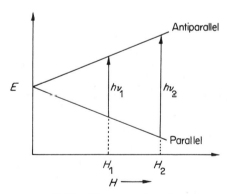

Figure 11.15. The variation in energy between the two spin states of an electron with the strength of the applied magnetic field

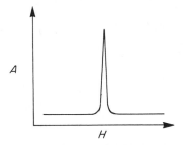

Figure 11.16. A simulated e.s.r. absorption spectrum

strength, and for a single electron would look like the simulated spectrum in Figure 11.16, although in practice it is usually the first derivative (dA/dH) of this line which is actually recorded.

The single absorption may be split if the nucleus with which the electron is associated possesses a magnetic moment. In such a case the nuclear magnetic moment takes up one or other of the allowed positions with respect to the direction of the applied field. If we consider a simple case where the nuclear spin quantum number is $\frac{1}{2}$ so that there are two allowed positions of the nuclear magnetic moment in the applied field, we have the situation shown in Figure 11.17, which produces a spectrum with two peaks as in Figure 11.18.

If the nuclear spins of the nuclei involved in a complex are known, and if the electron is associated with these nuclei (one or more), this association has its consequences on the nature of the e.s.r. spectrum.

Experimental studies have shown that unpaired electrons do associate not only with the central metal ions but also with the ligand atoms. Thus there is much evidence for molecular orbital formation in metal–ligand complexes. When internuclear distances are considered, and in particular

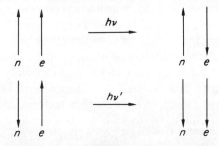

Figure 11.17. The effect of nuclear spin upon changing electronic spin states

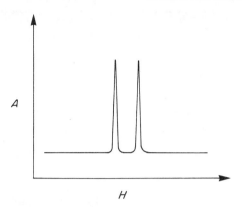

Figure 11.18. The e.s.r. spectrum for an electron in the presence of a nucleus of spin $\frac{1}{2}$

those between ligand atoms (i.e. N of NH_3) and the central metal atoms, it is obvious from their radial functions that there is considerable overlap.

A full molecular orbital treatment of such compounds incorporates all the advantages of the crystal field theory (then referred to as ligand field theory in which point charges are replaced by actual ligands) but is very complex. A brief outline of the results of such theory is presented for various types of metal–ligand compounds. The outline of molecular orbital theory as applied to octahedral complexes, only, is given. The similar treatments of square planar and tetrahedral complexes are omitted, but in general are as successful in the interpretation of the bonding, spectra and magnetic properties as that for octahedral species.

Molecular Orbital Theory of Octahedral Complexes

Symmetry theory shows that for octahedral complexes it is possible to make use of the e_g orbitals of the central metal atom along with the vacant s and three p orbitals in molecular orbital formation. The s and p orbitals normally have principal quantum number values which are greater by one than that of the d orbitals. For example, if we consider a first row transition element the $3d$ orbitals will be used together with the $4s$ and $4p$ orbitals. The six ligand orbitals may be arranged to form six group orbitals which have the correct symmetries to form effective molecular orbitals with the orbitals of the metal atom. In this way *six* sigma-bonding levels are produced and the corresponding number of antibonding orbitals. The three t_{2g} orbitals cannot take part in such sigma bonding and must therefore remain as non-bonding orbitals.

Molecules and Ions Containing Transition Elements

The six bonding orbitals are shown in Figure 11.19 as overlap diagrams, with the ligand orbitals being represented by the part which is directly concerned in overlap with the central metal atom.

In Figure 11.19, (a) represents the overlap of the metal s orbital with the appropriate ligand group orbital. This is totally symmetric (a_{1g}) with respect to octahedral symmetry. Orbitals (b), (c) and (d) are those which involve the metal p orbitals. They are less symmetrical than (a), but together form a triply degenerate set (t_{1u}). Orbitals (e) and (f) are also degenerate (e_g)

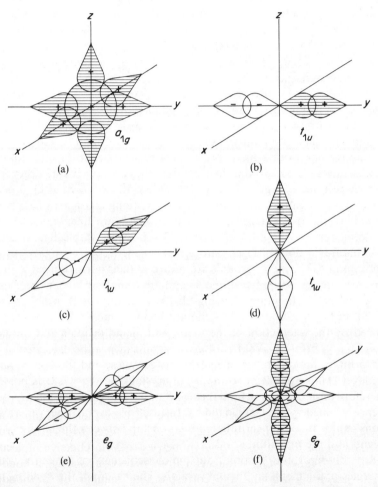

Figure 11.19. Representation of the six bonding molecular orbitals in an octahedral ML_6 complex

and are formed from the $d_{x^2-y^2}$ and d_{z^2} metal orbitals respectively, together with the appropriate ligand group orbitals.

An equivalent description of the bonding of an octahedral complex in terms of localized electron pair bonds is given by taking suitable linear combinations of the a_{1g}, t_{1u} and e_g molecular orbitals. The localized orbitals may then be described as being constructed from individual ligand orbitals and sp^3d^2 hybridized metal orbitals.

The six bonding molecular orbitals are not all of the same energy. The symmetric bonding orbital (a) is the lowest in energy since the orbital is bonding in all metal–ligand directions and is also ligand–ligand bonding. The triply degenerate orbitals (b), (c) and (d) are metal–ligand bonding and will be higher in energy than (a). The doubly degenerate orbitals (e) and (f) are of higher energy still since they contain appreciable ligand–ligand antibonding character.

It is possible to construct an energy diagram which shows the relative energies of metal, ligand and metal–ligand complex orbitals. In general, the ligand orbitals are at lower energies than the metal orbitals. That is, it is more difficult to remove (ionize) an electron from a ligand than it is to remove one from a metal atom (or ion). The ligand orbitals may be represented as group orbitals in which case they are not six-fold degenerate but share three energies. The completely symmetric orbital (a_{1g}) is the most stable because of its ligand–ligand bonding character. This is followed by the triply degenerate set (t_{1u}) which are ligand–ligand non-bonding (or very slightly antibonding). The least stable ligand group orbitals are the doubly degenerate (e_g) set which, overall, are ligand–ligand antibonding. The metal orbitals are shown in their normal order with the $3d$ set being lowest and with the $4s$ and $4p$ orbitals at higher energies.

A qualitative diagram showing all these energies is shown in Figure 11.20. Figure 11.20 also shows the antibonding molecular orbitals produced by the interaction of the metal and ligand orbitals and the non-bonding t_{2g} orbitals. What is of great significance is the existence of the e_g* orbitals, the difference in energy between these and the non-bonding t_{2g} level being equivalent to the Δ_0 of crystal field theory. It is possible therefore for molecular orbital theory to explain all the phenomena which were explained by crystal field theory. In addition, several other generalizations can be made concerning the relationship between the nature of the ligand and the magnitude of Δ_0.

It is of importance to point out the consequences of the difference in energies of the metal and ligand orbitals. The nature of the bonding and antibonding orbitals is strongly dependent upon the difference in energy between the contributing atomic orbitals (Section 10.5). Taking the general

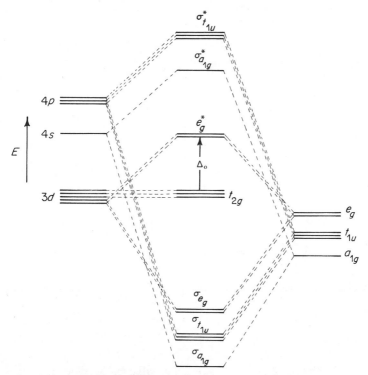

Figure 11.20. The molecular orbitals of an octahedral complex

case where the ligand orbitals are the more stable we may conclude that the bonding orbitals will be mainly of ligand orbital character and that the antibonding orbitals will be mainly of metal orbital character.

Relative to the energies of the metal orbitals, those of the ligand orbitals will be higher, the greater the electron donating power of the ligands. Since the ammonia molecule is a better electron donor than, say, the fluoride ion, we may conclude that the ammonia orbitals will contribute less to the bonding orbitals of a metal–ammonia complex than the fluoride orbitals will contribute to those of a metal–fluoride complex. This situation is illustrated diagrammatically in Figure 11.21. We may take Co^{III} to illustrate the effect of ammonia molecules and fluoride ions as ligands. The Co^{III} ion is a d^6 species and together with the twelve electrons supplied by the six ligands there are eighteen electrons to be distributed in the molecular orbitals. Feeding the electrons into the lowest levels in the case of the ammonia molecule we get the configuration,

$$(a_{1g})^2 (t_{1u})^6 (e_g)^4 (t_{2g})^6$$

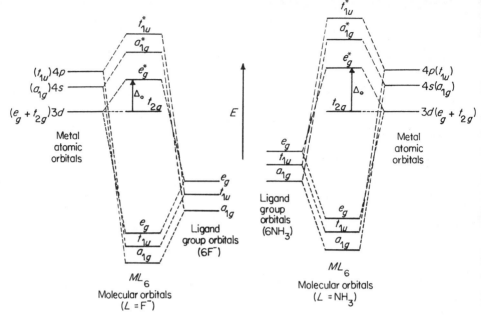

Figure 11.21. The molecular orbitals of MF_6 and $M(NH_3)_6$ complexes

for the $Co(NH_3)_6^{3+}$ complex since the value of Δ_0 is greater than the energy of electron pairing. This value (Δ_0) is decreased when the ammonia ligands are replaced by fluoride ions. The fluoride orbitals are of considerably lower energy and the bonding is more asymmetric in that the bonding orbitals are much more similar to the ligand orbitals in this case. In consequence, Δ_0 is less than in the hexammine cobalt(III) ion, and it is then possible for a spin-free configuration to be the more stable and we have:

$$(a_{1g})^2 (t_{1u})^6 (e_g)^4 (t_{2g})^4 (e_g^*)^2$$

as the electronic configuration of the CoF_6^{3-} complex.

The above considerations explain why the fluoride ion exerts a lower *ligand field* than does the ammonia molecule. In general, one might expect the spectrochemical series to be the order of ligands in terms of their increasing powers of electron donation. A good electron donor would be capable of good coordinate bond formation with a metal, with the metal orbitals making a reasonable contribution to the bonding orbitals.

Molecules and Ions Containing Transition Elements

Without doubt the extent of electron donation in the formation of sigma bonds is very important in the determination of Δ_0. The greater the tendency for donation, the greater should be the difference in the energies of the e_g^* and t_{2g} levels. The absorption spectra of complexes are influenced by the change in the value of Δ_0.

Sigma electron donation is, however, not the only factor and is unable, for instance, to explain why the iodide ion exerts a smaller ligand field than the fluoride ion, nor why the cyanide ion, carbon monoxide and ethylene are at the top of the spectrochemical series, none of which are very good electron donors. Obviously there are other factors in operation which have their influences at opposite ends of the spectrochemical series.

What we have ignored up to now is the possibility of pi bond formation. Symmetry theory shows that π-type overlap is possible between ligand p or d non-bonding or π^* antibonding orbitals and the t_{2g} metal orbitals, the latter being non-bonding in solely sigma bonded complexes.

(a) *Pi Bond Formation using Ligand Non-Bonding Orbitals*

Consider the case of the iodide ion, I^-, which has two non-bonding p orbitals perpendicular to the metal–ligand bond directions in MI_6 complexes. A typical overlap may occur between the d_{xy} orbital and the p orbitals of four of the iodide ligands, as shown in Figure 11.22. Similar overlaps are possible which involve the d_{xz} and d_{yz} orbitals.

If such interaction takes place this will have the effect shown in Figure 11.23. The pi bonding combinations will be at lower energy than the

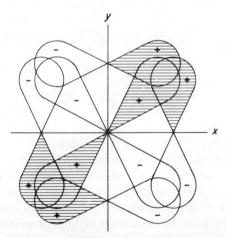

Figure 11.22. Overlap of the d_{xy} orbital of a central metal atom and four p orbitals of ligand groups

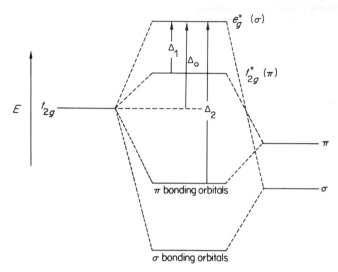

Figure 11.23. Effect of pi bond formation upon the magnitude of Δ

ligand pi orbitals and of great importance is that the energy of the $\pi^*(t_{2g}^*)$ antibonding combination will be greater than that of the original t_{2g} orbitals. This has the effect of *decreasing* the effective value of Δ since this will now be the difference in energy between the e_g^* and $\pi^*_{t_{2g}}$ orbitals, Δ_1. If the t_{2g}^* orbitals are vacant, then the ligand field splitting is given by the difference in energy between the e_g^* and the t_{2g} bonding levels, Δ_2.

It is therefore possible for iodide ion to exert a smaller ligand field than fluoride ion even though it is a better electron donor. Its pi interaction outweighs the sigma interaction, and Δ for the iodide ion is very low. Such ligand to metal 'donation' is important for any donor atom with p lone pairs available. It can occur even if the metal possesses a t_{2g}^6 configuration, but can only be regarded as a real donation if the t_{2g} levels do not have their maximum complement of electrons (six). Similar interaction can occur between the t_{2g} metal orbitals and the normally vacant d orbitals of the ligand atoms. For example, the ligands triphenylphosphine sulphide, Ph_3PS and the substituted arsines, R_3As can make use of the d orbitals of the sulphur and arsenic atoms respectively.

When interaction between the t_{2g} metal orbitals and the vacant d orbitals of the ligands occurs the effect is to stabilize the electrons originally present in the t_{2g} orbitals. They occupy the bonding level (t_{2g}) and this has the effect of increasing the value of Δ.

With ligands which possess occupied p orbitals and vacant d orbitals,

Molecules and Ions Containing Transition Elements

both of the above effects can occur. The two effects have opposing influences upon the value of Δ.

(b) *Pi Bond Formation using Ligand Antibonding Pi Orbitals*

The cyanide ion, CN^-, and carbon monoxide, CO, are isoelectronic, neither of them being good pi donors. They are, however, at the top of the spectrochemical series (i.e. exert large ligand fields and cause Δ to be large). The bonding of the carbon monoxide molecule has been dealt with in detail (Section 7.8) and that in the cyanide ion is very similar in that in both cases there is a lone pair orbital on the carbon atom (probably an *sp* hybrid type of orbital) which may be used in sigma bond formation with a suitable metal atom. There are some significant differences between the two ligands in respect of their different charges. The cyanide ion forms stable complexes with the positive oxidation states of a metal whereas the neutral carbon monoxide prefers either low oxidation states or more commonly the zero oxidation state of a metal. Both of these tendencies are understandable in terms of the Electroneutrality Principle (Section 11.3).

As far as the bonding is concerned there are great similarities. Both ligands can take part in sigma bond formation in the usual way and if only sigma bonding was allowed, then the ligand field effects would be small since neither of the ligands are good electron donors. They both, however, have pi bonds and consequently they have vacant π^* antibonding orbitals.

These latter have just the correct symmetry properties for overlap with the metal t_{2g} orbitals. The formation of a bonding orbital between a π^*

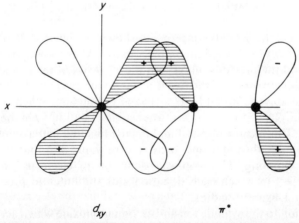

Figure 11.24. The overlap of a d_{xy} orbital of a metal atom with a π^* orbital of a cyanide ion (or a carbon monoxide molecule)

orbital and a d_{xy} orbital is shown in Figure 11.24. The metal–carbon–oxygen or metal–carbon–nitrogen atoms are collinear in consequence of such bonding, as is found by observation.

It is possible in an octahedral complex such as $Co(CN)_6^{3-}$ or $Cr(CO)_6$ for group orbitals produced from the π^* orbitals of the ligands to overlap with the metal orbitals. The situation for the d_{xy} orbital is shown in Figure 11.25. The effect of such (metal–ligand) pi bond formation upon

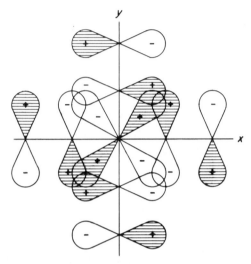

Figure 11.25. The overlap of the d_{xy} orbital of a metal ion with four π^* ligand orbitals

the energies of the orbitals concerned is shown in Figure 11.26 which also shows the effect of pi bonding on the value of Δ. The difference between this case and the previous one is that the π^* levels of the ligands do not contain any electrons and the t_{2g} electrons which are present in the metal are considerably stabilized by this type of pi bonding. When such interaction takes place, it is equivalent to electron donation from the metal to the ligand. This has the effect of making the ligand a better sigma donor. The pi bonding, then, not only increases the value of Δ, but strengthens the sigma bonding. This two-way donation of electrons produces a *synergic* effect in which each donation aids the other to produce very strong metal–ligand bonding.

In addition to the cyanide ion and carbon monoxide which act as sigma donors and pi acceptors, the ethylene molecule also forms very stable complexes with some metals, although the bonding is somewhat different.

Molecules and Ions Containing Transition Elements

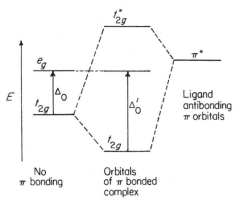

Figure 11.26. The effect of metal–ligand pi interaction upon Δ

In the complex $[C_2H_4Pt^{II}Cl_3]^-$, the platinum(II) ion is surrounded by the four ligands in a square planar mode with the carbon–carbon axis of the ethylene molecule at right angles to the square plane, as in Figure 11.27. The platinum(II) ion is a d^8 species for which square planar shapes

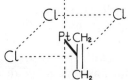

Figure 11.27. The structure of the $[C_2H_4Pt^{II}Cl_3]^-$ complex

are expected due to a strong Jahn–Teller effect. The ethylene molecule cannot enter into sigma bonding in the same way as normal sigma donors, since it has no suitable orbitals collinear with its molecular axis. We may consider the spatial arrangements and symmetries of the platinum orbitals and the ethylene orbitals, having prior knowledge of the shape of the complex. It can be seen from Figure 11.28 that they have the correct symmetries for molecular orbital formation. The 'sigma' bond is composed of an overlap of one of the metal orbitals with the pi bonding orbital of the ethylene molecule. As such this would imply an ethylene to platinum donation. Platinum to ethylene donation may occur by the overlap of the d_{xz} orbital and the π^* orbital of the ethylene molecule. Thus we have the synergic effect producing a stable bonding situation. There is adequate experimental justification for these synergic sigma–pi bonds. Some examples follow.

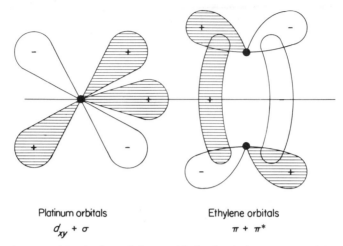

Figure 11.28. The d_{xy} and sigma orbitals of a platinum atom and the π, π^* orbitals of the ethylene molecule

(i) In metal carbonyls the metal to carbon bond distance is found in general to be shorter than that expected for a single covalent bond. The carbon–oxygen distance is generally greater than that in carbon monoxide itself. These facts are consistent with the metal–carbon bond order being greater than one, that of the C—O bond being less than three. This is to be expected if the carbon monoxide is involved in both sigma and pi bonding with the metal and the pi bonding involves donation of electrons to the antibonding (with respect to C—O) π^* orbital of the ligand.

(ii) In complexes involving ethylene as ligand it has been observed that the carbon–carbon distance in the ligand is in the range 0·140 to 0·147 nm. This is considerably longer than the corresponding distance in free ethylene (0·134 nm) and is consistent with there being occupancy of the π^* orbital.

(iii) The molecule of carbon disulphide, CS_2, is expected to be linear (cf. CO_2) in its electronic ground state. It can behave as a ligand and an example of a complex in which it does is shown in Figure 11.29.

The ligand has a bond angle of 136° and the C—S bond lengths on average are 0·163 nm. The length in the ground state is 0·155 nm. It would be expected from the Walsh diagram for AB_2 molecules (Figure 8.31) that the first excited state of CS_2 would be bent. Such a prediction concurs with observation that the first excited state is bent with an angle of 135° in which the average bond length is 0·164 nm. This geometry is almost exactly equivalent to that of the molecule when acting as a ligand. If electron density is donated to the platinum atom from the pi bonding

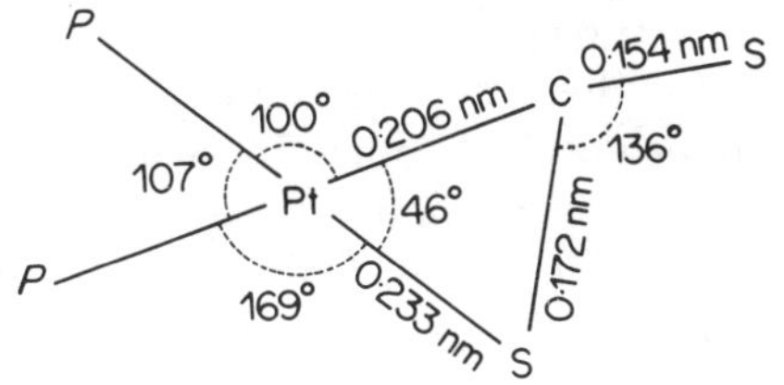

Molecules and Ions Containing Transition Elements

Figure 11.29. The structure of $P_2\text{PtCS}_2$ where $P = \text{Ph}_3\text{P}$ (triphenylphosphine)

levels of the carbon disulphide and received from the platinum atom d orbitals by the antibonding π^* orbitals of the CS_2, then this is equivalent to a $\pi \to \pi^*$ transition. It would be expected that in addition to forming a stable synergic bond with the platinum atom, the carbon disulphide molecule would have a geometry equivalent to its first excited state. This seems to be the case.

(iv) The nitrogen molecule is isoelectronic with that of carbon monoxide, and in the past there has been speculation as to whether nitrogen molecules could be coordinated to a metal atom in a manner similar to carbon monoxide. We may take as a guide to such a possibility the lowest energy $n \to \pi^*$ transition of the two molecules. This is justifiable in view of what is known about the synergic effect. The higher the transition energy the less probable would be the possibility of chemical binding offsetting it to produce a stable compound.

The $n \to \pi^*$ transition energy of CO is 6 ev and that of N_2 is 7·3 ev. It would seem less likely therefore that nitrogen molecules would form stable complexes and until recently none were observed. There are many complexes involving carbon monoxide and, in December 1965, the first complex ion involving nitrogen as a ligand was reported. This was nitrogenopentammine ruthenium(II) ion, $[\text{Ru}(\text{NH}_3)_5\text{N}_2]^{2+}$ formed in association with halide ions or the borofluoride anion. The Ru—N—N bonding appears to be linear with a nitrogen–nitrogen bond length of 0·112 nm. This is to be expected if the nitrogen is bonding in a similar manner to carbon monoxide.

11.5 Summary

Both the crystal field theory and the ligand field molecular orbital theory can reasonably explain the stereochemical, spectroscopic, magnetic and other properties of transition metal compounds. The latter is much the more complicated of the two theories but is also the more successful in that it can give an idea of the bonding in such compounds in terms of overlapping orbitals. It must be pointed out that the best approach to these theories at the moment is to make first-order calculations and then fit to these the experimental (say, spectroscopic) measurements of Δ, which allows considerable prediction of other properties. Neither theory allows an *a priori* calculation of a quantity such as Δ but they both are very useful in an empirical and qualitative sense.

Problems

11.1 Calculate some appropriate crystal field stabilization energies to derive justification for the fact that no spin paired tetrahedral complexes exist.

11.2 Apply the Jahn–Teller effect to asymmetrically occupied t_{2g} orbitals and list those d electron configurations which should lead to distortion of the complex in which they are situated. How would the degeneracy of the t_{2g} orbitals split?

11.3 What would be the localized bond descriptions of the bonding in ML_6 (octahedral), ML_4 (square planar) and ML_4 (tetrahedral) complexes?

11.4 Complexes of transition metals are far more stable than complexes formed by the main group elements. Discuss possible reasons why this should be so.

11.5 Predict the relative sizes of the $+3$ oxidation states of the first row of transition elements.

11.6 From the results of problem (11.5), predict the variation in hydration energies of these ions and the lattice energies of their fluorides.

11.7 Consider the inclusion of an electron pairing energy term in the calculations of crystal field stabilization energies.

11.8 The compound VCl_4 has a single absorption peak at 1100 nm. Calculate the absorption peak in the complex ion VCl_6^{2-}. Assign the two absorptions to particular transitions.

11.9 What d electron configurations would be subject to the Jahn–Teller effect if the metal atom were in a tetrahedral environment?

11.10 $Co(H_2O)_6^{2+}$ has three unpaired electrons and is stable in aqueous solution. In the presence of ammonia it is easily oxidized to $Co(NH_3)_6^{3+}$. Explain.

Appendix I

Slater's Rules for the Calculation of Effective Nuclear Charge

1. The orbitals are divided into the following groups:

    (a) $1s$
    (b) $2s, 2p$
    (c) $3s, 3p$
    (d) $3d$
    (e) $4s, 4p$
    (f) $4d$
    (g) $4f$, etc.

2. A contribution to the screening constant, S, is calculated for the electrons in each appropriate group:

    (a) nothing from any electron outside the group being considered,
    (b) an amount, 0·35, for each other electron in the group being considered (take 0·30 if the group being considered is the $1s$), and
    (c) an amount, 0·85, for each electron in the next lower group, if it is an sp group, and an amount 1·00 for all other electrons nearer the nucleus. For d and f groups an amount 1·00 for each electron should be taken.

Example

The oxygen atom has eight electrons distributed as $1s^2\,2s^2\,2p^4$. The nuclear charge, Z (eight for oxygen), would be effectively reduced by the eight electrons which would amount to a screening constant, S, made up from $(6 \times 0.35) + (2 \times 0.85) = 3.8$. The effective nuclear charge is thus $Z - S = 8 - 3.8 = 4.2$ for any additional electrons. Each electron in the $2s, 2p$ group would experience an effective nuclear charge of $8.0 - (5 \times 0.35) + (2 \times 0.85) = 4.55$.

Appendix II

The Successive Ionization Potentials (eV) of the Elements of the First Short Period

Element	1	2	3	4	5	6	7	8
Li	5.39	75·6	122·4					
Be	9·32	18·2	153·5	217·7				
B	8·3	25·1	37·75	259·3	340·1			
C	11·26	24·38	47·26	67·48	392·0	487		
N	14·54	29·6	47·43	77·45	97·86	556	663	
O	13·64	35·15	54·93	77·39	113·9	138	739	867
F	17·42	35·0	62·65	86·23	114·2	157	185	953
Ne	21·56	41·07	64·0	97·16	126·4	159	207	240

Appendix III

The Order of Filling of the Orbitals of some Homonuclear Diatomic Molecules

It was pointed out in Section 7.3 that the strength of a chemical bond (as measured by the stabilization of the bonding electrons with respect to the separated atoms) depends on the value of the overlap integral S (equation 7.44).

It is of interest to inquire into the way in which S varies with the internuclear distance in homonuclear diatomic molecules.

Mulliken and his co-workers (*Journal of Chemical Physics*, **17**, 1248–1267, 1949) have calculated values of the overlap integrals for various cases. The values of S are those calculated for Slater orbitals. These are a good approximation to the non-hydrogenic self-consistent-field orbitals. They have the forms:

$$\psi_{2s} = N_{2s} r e^{-Z_{\text{eff}} r / 2}$$

$$\psi_{2p_x} = N_{2p_x} x e^{-Z_{\text{eff}} r / 2}$$

$$\psi_{2p_y} = N_{2p_y} y e^{-Z_{\text{eff}} r / 2}$$

$$\psi_{2p_z} = N_{2p_z} z e^{-Z_{\text{eff}} r / 2}$$

where the N's are normalizing factors, Z_{eff} is the effective nuclear charge (calculated by Slater's Rules, Appendix I), and r is the distance from the nucleus in atomic units (0·053 nm). The basis of the atomic unit of length is that 0·053 nm is the distance from the hydrogen nucleus where the radial function is a maximum. To express a distance in atomic units the distance in nanometres is divided by 0·053 to give a dimensionless number.

The variations of the overlap integrals for (a) 2s–2s, sigma overlap, (b) 2p–2p, sigma overlap, (c) 2p–2p, pi overlap and (d) *sp–sp*, hybrid sigma overlap with the parameter ρ are shown in the figure.

The parameter ρ is given by $(Z_{\text{eff}} R)/2$ where R is the internuclear separation of the two atoms in a homonuclear diatomic molecule (in

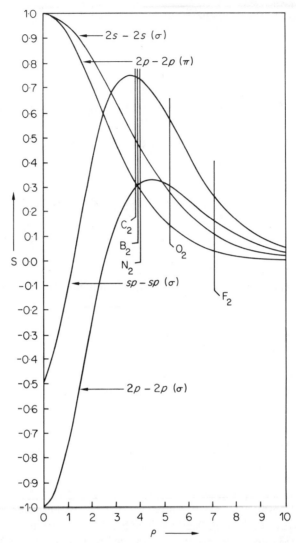

Figure: Plots of the overlap integrals versus the parameter ρ for (a) 2s–2s, sigma overlap, (b) 2p–2p, sigma overlap, (c) 2p–2p, pi overlap and (d) sp–sp hybrid sigma overlap

atomic units). The values of ρ for the equilibrium bond lengths of the molecules B_2, C_2, N_2, O_2 and F_2 are indicated in the figure.

The graphs of S versus ρ can be understood by a consideration of the

The Order of Filling of the Orbitals

spatial distributions of the participating s and p orbitals (Figures 5.5 and 5.6).

(a) When two $2s$ orbitals overlap it would be expected that as the internuclear separation, R, decreases S would increase to a maximum value of $1\cdot0$ when R (and ρ) becomes zero, i.e. when the nuclei coincide and the two orbitals become completely superimposed.

(b) In this case two $2p$ orbitals overlap in a sigma manner and at large R (or ρ) values the overlap integral is positive (all S values become zero at $R = \infty$) and eventually passes through a maximum at $\rho \sim 4\cdot 6$. At values smaller than this there is an ever increasing contribution to S of positive–negative overlap which decreases the value of S. In the position of complete superimposition ($\rho = 0$) there is complete positive–negative overlap and S has the value of $-1\cdot 0$.

(c) When two $2p$ orbitals overlap in a pi manner there is complete positive–positive, negative–negative overlap when ρ is zero so that S increases as ρ decreases and the limiting value of S is unity.

(d) When two sp hybrid orbitals (directed towards each other) overlap there is positive overlap at large ρ values but as in case (b) there comes a point where S passes through a maximum (at $\rho \sim 3\cdot 6$) due to the increasing amount of positive–negative overlap as ρ decreases. At the position, $\rho = 0$, the positive–negative overlap is a maximum but, since the negative lobes are smaller than the positive ones, the limiting value of S is only $-0\cdot 5$.

In the latter case, however, the maximum value of S ($0\cdot 76$) is considerably greater than that in the case of $2p$–$2p$ sigma overlap ($0\cdot 332$). Since bond stability depends on S this is a very good justification for the operation of hybridization. The extent to which hybridization occurs depends on the energy gap between the s and p orbitals and this gap does vary from element to element (Figure 8.18). The molecules B_2 and C_2 are composed of atoms where the s–p gap is small and it would be expected that the σ_{2s}, σ_{2s}^*, σ_{2p} and σ_{2p}^* orbitals would participate considerably in hybridization. The symbols are, however, still retained for most purposes. The s–p energy gap increases considerably in nitrogen, oxygen and fluorine and it would be expected that the σ_{2s}, σ_{2s}^*, σ_{2p} and σ_{2p}^* orbitals would be composed mainly of pure s or p contributions respectively.

In the molecules B_2, C_2 and N_2 the equilibrium values are such that the overlap integrals for the π orbitals are quite considerable, unlike the situation in F_2 where the value is negligible. It would be expected that the pi bonding orbitals would have a reasonable stability but even if hybridization is discounted (hybridization increases S for the σ bonds), it would appear that the σ_{2p} bonding orbital should be of lower energy (because $S\sigma > S\pi$). With the inclusion of hybridization there would seem to be no doubt that the σ_{2p} level should be more stable than the π_{2p} level.

There is, however, another factor to be considered. This is the interaction between levels of the same symmetry. The levels which can take part in this interaction are the σ_{2s}, σ_{2p} and σ_{2s}^* and σ_{2p}^* pairs. The σ_{2s} and σ_{2p} are both sigma and *gerade* in character while the σ_{2s}^* and σ_{2p}^* are both sigma and *ungerade* in character.

The interaction which takes place is very similar to that taking place between two atomic orbitals (one on each atom) of the correct symmetry in forming a bonding and antibonding orbital pair.

The interaction of the σ_{2s} and σ_{2p} orbitals produces a stabilizing of the σ_{2s} level and a corresponding destabilizing of the σ_{2p} level. Similarly the σ_{2s}^* level is stabilized and the σ_{2p}^* level is destabilized by the interaction between the σ_{2s}^* and σ_{2p}^* levels.

The extent of this interaction depends on the closeness in energy of the two participating levels. The closer together the original levels are in energy the greater is the interaction. In consequence, the interaction decreases along the series B_2, C_2, N_2, O_2 and F_2, being very important in the first three molecules. It is this interaction which destabilizes the σ_{2p} level (in particular) and in B_2, C_2 and N_2 causes it to have a higher energy than the corresponding π_{2q} level. In O_2 and F_2 the s–p energy gap is sufficient to prevent the interaction from having such a significant effect. The interaction still takes place but is not sufficient to raise the σ_{2p} level above that of the π_{2p}.

References for Further Reading

Key references for those who wish to read further about the theories presented in the book are listed below:

P. G. Dickens and J. W. Linnett, 'Electron correlation and its chemical consequences', *Quarterly Reviews*, **XI**, 291 (1957).
R. J. Gillespie and R. S. Nyholm, 'Inorganic stereochemistry', *Quarterly Reviews*, **XI**, 339 (1957).
G. W. A. Fowles, 'Lone pair electrons', *J. Chem. Ed.*, **34**, 187 (1957).
L. S. Bartell, 'Molecular geometry', *J. Chem. Ed.*, **45**, 754 (1968).
R. M. Gavin, Jr., 'Simplified molecular orbital approach to inorganic stereochemistry', *J. Chem. Ed.*, **46**, 413 (1969).
C. A. Coulson, 'The representation of simple molecules by molecular orbitals', *Quarterly Reviews*, **I**, 144 (1947).
J. A. Pople, 'The molecular orbital and equivalent orbital approach to molecular structure', *Quarterly Reviews*, **XI**, 273 (1957).
T. C. Waddington, 'Lattice energies and their significance in inorganic chemistry', *Advances in Inorganic and Radiochemistry*, **1**, 158 (1959).
A. L. Allred and E. G. Rochow, 'A scale of electronegativity based on electrostatic force', *Journal of Inorganic and Nuclear Chemistry*, **5**, 264–269 (1958).
L. N. Ferguson, 'Hydrogen bonding and the physical properties of substances', *J. Chem. Ed.*, **33**, 267 (1956).
A. D. Walsh, 'The electronic orbitals, shapes and spectra of polyatomic molecules', *J. Chem. Soc.*, **1953**, 2260–2331.
N. N. Greenwood, 'Principles of atomic orbitals', *Royal Institute of Chemistry Monographs for Teachers*, No. **8**.
P. J. Wheatley, *The Determination of Molecular Structure*, Oxford University Press, 1959.
J. W. Linnett, *Wave Mechanics and Valency*, Methuen, 1960.
J. W. Linnett, *The Electronic Structures of Molecules: A New Approach*, Methuen, 1964.
A. Liberles, *Introduction to Molecular Orbital Theory*, Holt, Rinehart and Winston, 1966.
J. N. Murrell, S. F. A. Kettle and J. M. Tedder, *Valence Theory*, Wiley, 1966.
L. E. Orgel, *An Introduction to Transition Metal Chemistry: Ligand Field Theory*, 2nd Edition, Methuen, 1966.
F. A. Cotton and G. A. Wilkinson, *Advanced Inorganic Chemistry*, 2nd Edition, Wiley, 1966.
W. E. Addison, *Structural Principles in Inorganic Compounds*, Longmans, 1961.
Banesh Hoffman, *The Strange Story of the Quantum*, Pelican Books, 1963.

Answers to Problems

1.1 (a) $1{\cdot}365 \times 10^{15}$ Hz; 45,400 cm^{-1}; $9{\cdot}05 \times 10^{-12}$ erg; $5{\cdot}7$ eV; 131 kcal.mole^{-1}; 547 kJ.mole^{-1}
(b) $4{\cdot}42 \times 10^{14}$ Hz; 14,700 cm^{-1}; $2{\cdot}925 \times 10^{-12}$ erg; $1{\cdot}84$ eV; $42{\cdot}4$ kcal.mole^{-1}; 177 kJ.mole^{-1}
(c) $1{\cdot}5 \times 10^{13}$ Hz; 500 cm^{-1}; $9{\cdot}94 \times 10^{-10}$ erg; $0{\cdot}062$ eV; $1{\cdot}44$ kcal.mole^{-1}; 6 kJ.mole^{-1}

1.2 $2{\cdot}3$ eV.atom^{-1} = 53 kcal.mole^{-1} = 222 kJ.mole^{-1}

2.5 $1{\cdot}14 \times 10^{-4}$ M; $31{\cdot}6\%$ transmission

2.6 3×10^3 l.mole^{-1} cm^{-1}

2.7 $8{\cdot}28$ kcal.mole^{-1} = $34{\cdot}6$ kJ.mole^{-1}

2.8 $h^2/4\pi^2 I = 10{\cdot}88$ cal.mole^{-1} = $7{\cdot}56 \times 10^{-16}$ erg.molecule^{-1}
$I = 14{\cdot}8 \times 10^{-40}$ erg.sec^2.molecule^{-1}
$= 14{\cdot}8 \times 10^{-40}$ g.cm^2.molecule^{-1}
(An erg is a cm.dyne; the dyne is mass × acceleration, i.e. it has units g.cm.sec^{-2}. A cm.dyne is thus g.cm^2.sec^{-2} and erg.sec^2 is equivalent to g.cm^2.)
$\mu = 11{\cdot}4 \times 10^{-24}$ g.molecule^{-1}
and $r = \left(\dfrac{I}{\mu}\right)^{1/2}$
$= 1{\cdot}14 \times 10^{-8}$ cm
$= 0{\cdot}114$ nm (the internuclear distance)

2.9 313 nm = 92 kcal.mole^{-1} = 384 kJ.mole^{-1}

3.2 Uncertainty in position of car $\sim 2{\cdot}2 \times 10^{-35}$ cm (very small in comparison with size of car), uncertainty in position of electron $\sim 0{\cdot}7$ nm (this is larger than most atoms).

4.1 $E = \dfrac{h^2}{8m}\left[\dfrac{n_x^2}{l_x^2} + \dfrac{n_y^2}{l_y^2} + \dfrac{n_z^2}{l_z^2}\right]$

4.2 (a) $l_x = l_y \neq l_z$
(b) $l_x = l_y = l_z$

Answers to Problems

Energy level	Quantum number values n_x	n_y	n_z	E in units of $h^2/8m$ (a)	E in units of $h^2/8m$ (b)
(1)	1	1	1	$\dfrac{2}{l_x^2}+\dfrac{1}{l_z^2}$	$\dfrac{3}{l_x^2}$
(2)	2	1	1	$\dfrac{5}{l_x^2}+\dfrac{1}{l_z^2}$	$\dfrac{6}{l_x^2}$
(3)	1	2	1	$\dfrac{5}{l_x^2}+\dfrac{1}{l_z^2}$	$\dfrac{6}{l_x^2}$
(4)	1	1	2	$\dfrac{2}{l_x^2}+\dfrac{4}{l_z^2}$	$\dfrac{6}{l_x^2}$
(5)	1	2	2	$\dfrac{5}{l_x^2}+\dfrac{4}{l_z^2}$	$\dfrac{9}{l_x^2}$
(6)	2	1	2	$\dfrac{5}{l_x^2}+\dfrac{4}{l_z^2}$	$\dfrac{9}{l_x^2}$
(7)	2	2	1	$\dfrac{8}{l_x^2}+\dfrac{1}{l_z^2}$	$\dfrac{9}{l_x^2}$
(8)	2	2	2	$\dfrac{8}{l_x^2}+\dfrac{4}{l_z^2}$	$\dfrac{12}{l_x^2}$

In case (a) levels (2) and (3) are degenerate,
    and levels (5) and (6) are degenerate.
In case (b) levels (2), (3) and (4) are degenerate,
    as are levels (5), (6) and (7).
Note that an increase in symmetry increases the extent of degeneracy. This is generally true.

5.1   (a) $9{\cdot}104 \times 10^{-28}$ g
     (b) $9{\cdot}1091 \times 10^{-28}$ g
     (c) $9{\cdot}1091 \times 10^{-28}$ g

5.6   The energy levels in a hydrogen-like atom are given by equation (5.4). The ionization potential of the hydrogen atom is obtained by putting $Z = 1$ and $n = 1$ in the equation: i.e.

$$I_\text{H} = -E_{n=1} = \frac{2\pi^2 me^4}{h^2}$$

Considering sodium to be a hydrogen-like atom, its first ionization potential is obtained by putting $Z = Z_{\text{effective}}$ and $n = 3$ (for the 3s electron) in equation (5.4): i.e.

$$I_\text{Na} = -E_{n=3} = \frac{2\pi^2 me^4 Z_{\text{eff}}^2}{9h^2}$$

The ratio of the two potentials is given by:

$$I_{Na}/I_H = Z_{eff}^2/9 = 2/5$$

Thus Z_{eff} for the $3s$ electron of the sodium atom is given by $(18/5)^{1/2} = \underline{1\cdot 9}$
Thus the $1s^2\,2s^2\,2p^6$ configuration reduces the nuclear charge of 11 to an effective one of $1\cdot 9$ as far as the experience of the $3s$ electron is concerned.

7.1 $\int_0^\infty \psi_S \psi_{AS}\,d\tau = \int_0^\infty (\psi_A + \psi_B)(\psi_A - \psi_B)\,d\tau$

$= \int_0^\infty (\psi_A^2 - \psi_B^2)\,d\tau = \int_0^\infty \psi_A^2\,d\tau - \int_0^\infty \psi_B^2\,d\tau = 0$

Providing that the atomic orbitals ψ_A and ψ_B are separately normalized, the last two integrals become unity and the orthogonality condition is fulfilled.

10.2 615 kcal.mole^{-1}, 2570 kJ.mole^{-1}

10.3 2510 kcal.mole^{-1}, 10,500 kJ.mole^{-1}

10.4 $\Delta H_f\,(CaF_{2(c)}) = -279$ kcal.mole^{-1}, -1165 kJ.mole^{-1}
$\Delta H_f\,(SnO_{2(c)}) = +145$ kcal.mole^{-1}, $+335$ kJ.mole^{-1}

10.7 $\Delta H_f\,(CaO_{2(c)})$ (containing O$^-$ ions) $= -136$ kcal.mole^{-1}, -314 kJ.mole^{-1}

11.8 The transition in VCl$_4$ at 1100 nm may be assigned to the orbital change $e^1 \to t_2^1$. In VCl$_6^{2-}$ the octahedral field would reverse the order of the energy levels and a $t_{2g}^1 \to e_g^1$ transition becomes possible. The transition would be expected to occur at approximately $1100 \times 4/9 = 490$ nm.

Index

A_2B_6 molecules 217
Absorbance 22
Absorption spectra 14, 22, 171, 274, 286
Absorption spectrum of benzene vapour 22
Acceptor molecule 209, 219
Acetylene, C_2H_2 216
Amino radical, NH_2 145, 150, 171
Ammonia, NH_3 197
Angular dependence of d orbitals 71
Angular dependence of p orbitals 70
Angular dependence of s orbitals 69
Answers to numerical problems 318
Antibonding orbitals 117, 120
Antisymmetric overlap 115
Appearance potential 26
Arsines 304
Atomic emission spectroscopy 14, 81
Atomic number xi
Atomic orbitals 64
Atomic weight xii
Atomic weight scale xi
Aufbau principle 98
Axes of symmetry 151
Azide ion, N_3^- 175
Azobenzene 216, 234

B_2 molecule 125, 130
BH_2 radical 147, 170
Balmer series 15
Band theory 265
Bathochromic shift 222
Beer–Lambert Law 22
Bent bonds 214
Benzene, C_6H_6 228

Beryllium hydride, BeH_2 147, 170
Beryllium molecule, Be_2 125, 130
Bidentate ligand 273
Blackbody radiation 3, 11
Bohr frequency condition 16
Bohr model of the hydrogen atom 46
Bond angle 145, 149, 167, 174, 177, 185, 188, 197, 214, 218, 220
Bond dissociation energy 121, 212, 215, 217
Bond length 23, 212, 215, 217
Bond order 130
Bond pair, bond pair repulsions 149, 191
Bond pair, lone pair repulsions 149, 191, 197
Bonding orbitals 113, 120
Borazole, $B_3N_3H_6$ 220, 233
Borine radical, BH_3 219
Born–Haber thermochemical cycle 255
Born–Landé equation 254
Born–Mayer equation 255
Boron trifluoride, BF_3 191
Boundary surface 69, 71
Brackett series 19
Bridged molecules 217
Butadiene 221

C_2 molecule 127, 130
C_{2v} character table 161
Calcium fluoride, CaF_2 241
Canonical forms 235
Canonical forms of the nitrite ion 177
Carbonate ion, CO_3^{2-} 193
Carbon dioxide 175, 184
Carbon disulphide 308

Carbon monoxide   139, 141, 305
Cartesian coordinates   47
Centre of symmetry   151, 153
Change in free energy, ΔG 262
Change in heat content, ΔH 255
Charge correlation   89, 97
Charge correlation energy   96
Chelate ring   273
Chlorine trifluoride, ClF_3 200
Chlorine trioxide, ClO_3 204
Chromium hexacarbonyl, $Cr(CO)_6$ 306
Cobalt octacarbonyl, $Co_2(CO)_8$ 214
Coding of atomic orbitals   65
Coding for resultant orbital angular momentum, L   83
Coloured compounds   275
Complex   273
Compton effect   10
Conduction band   263
Continuous emission   14
Coordinate bond   194
Coordination number   251, 263
Coulomb integral, α 118
Coupling between l and s 82
Coupling between L and S 83
Covalent bonding   110
Covalent radii   138
Covalent solids   238, 247
Crystal field stabilization energy   290, 294
Crystal field theory   278
Cyclohexane   232
Cyclohexatriene   229

d orbitals   65
d-d transitions   286
d_ε orbitals   279
d_γ orbitals   279
De Broglie's equation   39
Debye force   240
Degeneracy   56
Delocalization energy of benzene   231
Dewar forms of benzene   235
Dialuminium hexabromide   218
Dialuminium hexachloride   261
Dialuminium hexamethyl   220
Diamagnetism   275
Diamond   247

Diatomic molecules   110, 124, 133
Diborane, B_2H_6 218
Dichloroiodide ion, ICl_2^- 174
Diffraction   11
Diimine, N_2H_2 216
Dipole moment   134
Disproportionation   259
Distorted octahedral ions   283
Doublet state   84

e_g orbitals   279
Effective nuclear charge   28, 77
Eigenfunction   50, 62
Eigenvalue   50, 62
Electrical conduction   263
Electromagnetic radiation   1, 2
Electromagnetic spectrum   1, 2
Electron   37
Electron affinity   256
Electron correlation   89
Electron density   51
Electron diffraction   37
Electron in a potential well   52, 57, 61
Electron interchange   91
Electron pair donor   142, 302
Electron pairing   89
Electron spin   81
Electron spin resonance   295
Electronegativity   108, 136, 138
Electroneutrality principle   294, 305
Electronic configurations of the elements   98, 108
Electronic dissociation energy   121
Electronic sub-levels   30
Electronic transitions   24, 171, 222, 226, 229, 275
Element   xi
Elements of symmetry   151, 161
Emission spectrum of hydrogen   14
Emission spectrum of mercury   86
Emission spectrum of neon   20
Emission spectrum of sodium   21, 84
Energies of atomic orbitals   79
Energies of molecular orbitals   139, 159, 166, 179, 180, 184, 188, 224, 231, 250
Energy levels   14, 46
Entropy change, ΔS 262
Ethane   209

Index

Ethylene 212
Ethylene as ligand 307
Exothermic reaction 255

f orbitals 65
First Law of Thermodynamics 256
Flame test for sodium 81
Fluorine molecule, F_2 129
Forces, atomic xiii, 47
  coulombic xiii, 47, 252
  intermolecular xiii, 238
  molecular xiii
  nuclear xiii
  valence xiii
Formaldehyde 215
Fourier analysis 44
Free energy 262
Frequency, v 1
Fundamental particles xi
Fundamental vibration frequency, w_0 23

Gamma rays 2
Graphite 248
Ground state 20
Group orbitals 154, 163
Group theory 151

Hamiltonian operator 50
Heat of formation 255
Helium discharge lamp 34
Helium molecule, He_2 124
Helium molecule-ion, He_2^+ 123
Heteronuclear diatomic molecules 133, 139
Hexamminecobalt(III) ion 302
Hexacyanocobalt(III) ion 306
Hexafluorocobalt(III) ion 302
Hexaquocobalt(II) ion 295
Hexaquocopper(II) ion 288
Hund's Rules 100, 109, 127
Horizontal plane of symmetry 152
Hybrid atomic orbitals 131
Hybridization 131
Hydrazine, N_2H_4 210
Hydrazobenzene 234
Hydration energy 275

Hydrides, group 4 242
  group 5 242
  group 6 242
  group 7 242
  triatomic 145
Hydrogen bonding 242
Hydrogen bonds, intermolecular 246
Hydrogen bonds, intramolecular 246
Hydrogen cyanide 187, 216
Hydrogen difluoride ion, HF_2^- 246
Hydrogen emission spectrum 14
Hydrogen fluoride 133
Hydrogen molecule 111, 123
Hydrogen molecule-ion, H_2^+ 111, 122
Hydrogen peroxide 211
Hydrogen sulphide 189
Hydroperoxy radical, HO_2 187
Homonuclear diatomic molecules 124

Identity element of symmetry 153
Indistinguishability of electrons 90
Infra-red radiation 2
Insulators 265
Interelectronic repulsion 123
Interference phenomena 11
Intermolecular forces 238
Inversion centre 151, 153
Iodate ion 205
Iodine 241
Iodine dichloride ion, ICl_2^- 200
Iodine pentafluoride 201
Ionic bonding 110, 238, 252
Ionic solid 238, 249
Ionization limit 20, 25
Ionization potential 10, 26, 31, 33, 73, 256, 261, 312
Isomerism 212
Isotopes of hydrogen xii
Isotopy xii

Jahn–Teller effect 287
Jahn–Teller theorem 286

Keesom force 240

Laporte Rule 87
Lattice 239
Lattice energy 252, 275
Lewis acid 219

L.C.A.O. method 112
Ligand 194
Ligand, bidentate 273
Ligand, monodentate 273
Ligand field 302
Ligand field theory 295
Line emission 14
Lithium molecule, Li_2 124, 130
Localized bonds 159
Lone pairs 175
Lone pair, lone pair repulsions 191
London force 240
Lyman series 17

Madelung constant 253
Magnesium chloride 258
Magnetic balance 277
Magnetic quantum number, m 63
Magnetic moment 81
Mass number xi
Mass spectrometer 26
Matter waves 39
Metallic bond 263
Metallic conduction 263
Metallic state 239, 265
Metals 238
Mercuric chloride 174
Mercury dimethyl 145, 170
Mercury vapour discharge lamp 6
Methane 196
Methylene, CH_2 148, 170
MgCl 258
$MgCl_3$ 261
m-hydroxybenzaldehyde 246
Microwaves 2
m-nitrophenol 246
Molar absorption coefficient, ε 22
Molecular absorption spectroscopy 21
Molecular orbital 111
Molecular orbital theory 111
Molecular orbitals of octahedral complexes 298
Molecular solids 238
Moment of inertia 23
Multicentre sigma bonds 202
Multiple bonds 203
Multiplicity 84
Multiplicity forbiddenness 86

$NaCl_2$ 258
Naphthalene 241
Neon molecule, Ne_2 130
Neutrons xi
Nickel carbonyl 142
Nitrate ion 194
Nitric oxide 139
Nitrite ion 175, 186
Nitrogen dioxide 175
Nitrogen molecule 128, 130, 216, 309
Nitrogen molecule-ion, N_2^+ 128, 130
Nitronium ion, NO_2^+ 175
Nitrosyl chloride, ONCl 175
Nitrous oxide 175, 187
Non-bonding electrons 197
Non-bonding orbitals 303
Normalization of wave functions 56
Normalizing factor 114, 116
n-type conduction 267
Nuclear magnetic moment 297
Nuclear spin 297

Octahedral complexes 274
Octahedral crystal field splitting, Δ_0 279
Octahedral molecules 192, 197
o-nitrophenol 245
Orbital angular momentum 81
Orbital angular momentum quantum number, l 63
Orbital penetration effects 76
Orbitals of linear AB_2 molecules 178
Orbitals of linear triatomic hydride molecules 153
Orbitals of bent AB_2 molecules 180
Orbitals of bent triatomic hydride molecules 161
Order of filling of orbitals in atoms 97
Order of filling of orbitals in diatomic molecules 126, 313
Orthophosphate ion 205
Orthosilicate ion 205
o-salicylaldehyde 245
Overlap integral 114, 131
Overlapping of atomic orbitals 112
Oxidation 271
Oxidation state 271
Oxygen difluoride 185
Oxygen molecule 129, 130

Index

Oxygen molecule-ion, O_2^+ 129, 130
Oxygenyl hexafluoroplatinum (V) 129
Ozone, O_3 175

Pairing energy 279
Paramagnetism 275
Paschen series 19
Pauli exclusion principle 91, 97
$p \rightarrow d$ promotion 195
Pauling electroneutrality principle 294, 305
Perchlorate ion 205
Perfect radiator 12
Periodic classification of the elements 97, 108
Permanganate ion 294
Pfund series 19
PH_2 radical 189
Phosphorus pentachloride 196
Photoconductivity 267
Photoelectric absorption 10
Photoelectric effect 6, 11
Photoelectron kinetic energy 7
Photoelectron spectroscopy 34
Photoelectrons 6
Photoionization 33
Photon 9
Photon theory 8, 10
p-hydroxybenzaldehyde 246
Physical states of matter 238
p-nitrophenol 246
Pi bond 128
Pi bonds in transition metal complexes 303, 305
Pi bonding, effect on Δ_0 304
Pi electron delocalization 221
Pi orbitals 126
Pi $\rightarrow$ pi* transitions 24, 222, 229
Planes of symmetry 152
Planck's constant 9, 43
Planck's equation 9, 16, 39
Polar coordinates 47
Polarizable ligands 294
Polyelectronic atoms 73
Polyenes 226
p-orbitals 65
Potential energy curves 121
Principal quantum number, n 63
Principle of indeterminacy 41, 42, 60

Probability theory 51
Probability waves 12
Proton xi
Psi, Ψ 48
Psi, Ψ, interpretations of 51
  limitations to nature and form of 52
Problems 1 13
  2 35
  3 45
  4 61
  5 79
  6 109
  7 143
  8 190
  9 237
  10 268
  11 310
p-type conduction 268
$Pt(C_2H_4)Cl_3^-$ 307
$Pt(Ph_3P)_2CS_2$ 309

Qualitative analysis 21
Quanta 5
Quantization of radiation 1
Quantization of electron energies 14, 56
Quantum defect 25
Quantum mechanics 46
Quantum number 55
  n (principal) 17, 63
  l (orbital angular momentum) 63, 81
  m (magnetic) 63
  s (spin angular momentum) 81
  j (resultant of l and s) 83
  L (resultant orbital angular momentum) 82
  S (resultant spin angular momentum) 82
  J (resultant of L and S) 83
  J (rotational) 23
  v (vibrational) 23
  nuclear spin 297
Quantum numbers, non-zero values 60
Quantum theory 5

Radial probability distribution function 67
Radial wave function 63, 66

Index

Radiation theories 11
Radio waves 2
Rayleigh's equation 5
Reduced mass 62
Reduction 271
References for further reading 317
Refractive index 1
Relativity theory 11
Resonance 177
Resonance integral, β 118
Rest mass of electron 11
Rotational energy 23
Rotational quantum number, J 23
$Ru(NH_3)_5N_2^{2+}$ 309
Rydberg constant 72
Rydberg series 25, 34

s orbitals 65
$s \rightarrow d$ promotion 195
$s \rightarrow p$ promotion 147
Schroedinger equation 47
Scintillation 37, 38
Self-consistent field method 74
Self-consistent field orbitals 74
Semiconduction 266
Semiconductor 266
Series 15
Sidgwick–Powell theory 144, 173, 191
Sigma bond 123
Sigma bond, multicentre 202
Sigma orbital 123
Singlet state 84, 92
Slater's Rules 139, 311
Sodium chloride 251
Sodium metal 6
Spatial distributions of sigma electron pairs 192
Spectrochemical series 288
Spectroscopic states 84, 85, 87
Spin angular momentum 81
Spin quantum number, s 81
Spin correlation 89, 97
Spin correlation energy 96
Spin forbiddenness 86
Spin free configurations 281
Spin paired configurations 281
Spin reversal 295
Square planar complexes 284
Standing waves 54

Stefan's Law 5
Stereochemistry 274
Stern–Gerlach experiment 81
Stoichiometry of compound formation 258
Sublimation energy 257
Successive ionization potentials
  of argon 27
  of neon 29
  of potassium 27
  of Li to Ne 312
  of Na to P 261
Sulphate ion 205
Sulphite ion 204
Sulphur dioxide 175
Sulphur hexafluoride 197
Sulphur trioxide 195
Symmetric overlap 112
Symmetry characters 155, 161, 163
Symmetry elements 151, 161
Symmetry theory 150
Synergic effect 306

t_{2g} orbitals 279
Tellurium dibromide 174
Tellurium tetrachloride 199
Term symbols 83
Tetrachlorocobalt(II) ion 295
Tetrachloroiron(III) ion 294
Tetrachlorovanadium(V) ion 294
Tetrachlorozinc(II) ion 294
Tetrafluoroborate ion 192
Tetragonal distortion 283
Tetrahedral complexes 291
Tetrahedral crystal field splitting, Δ_t 293
Tetrahedral molecules 192, 197
Tetraoxoiron(VI) ion 294
Thought experiment 41
Threshold frequency 7
Threshold technique 27, 34
Tin metal 266
Tin(II) chloride 174
Titanium tetrachloride 294
Transition elements 102, 271
Transition probability 86
Triatomic hydrides 145
Triatomic molecules 144

Index

Trigonally bipyramidal molecules 192, 197
Trigonally planar molecules 192
Triiodide ion 185
Trimethylamine 207, 219
Triphenylphosphine 309
Triphenylphosphine sulphide 304
Triplet state 85, 92
Trisilylamine 207, 219

Ultraviolet radiation 2
Uncertainty principle (Heisenberg) 41, 42, 60
Uncertainty principle, consequences of 45
Uncertainty principle, expression for 43
Units xiv
Uranyl ion 175

Valence bond theory 235
Valence state 146
Van der Waals forces 238
Vanadium tetrachloride 294
Variation principle 74
Velocity of electromagnetic radiation 1
Vertical plane of symmetry 152
Vibrational energy 23
Vibrational quantum number, v 23
Visible light 2

Walsh diagram for AB_2 molecules 184
  for AH_3 molecules 167
  for HAB molecules 188
Walsh theory 145
Walsh's rules 168
Wave equation 46
Wave function 47
Wave function, angular 63
  radial 63
  time dependent 47
Wave mechanics 46
Wave motion 1, 3, 11
Wavenumber 3
Wavelength, λ 1
Water molecule 144, 150
Water, liquid 244
Wien's Law 4
Work function 9

X-rays 2
Xenon difluoride 202
Xenon tetrafluoride 203
Xenon hexafluoride 203

Zeeman effect 88
Zero-point vibrational energy 23, 45, 121
Zirconium tetrachloride 294